Quantitative Research Methods in
POLITICAL SCIENCE

FIRST CANADIAN EDITION

ANDREA WAGNER
Carleton University

Kendall Hunt
publishing company

Cover image © Shutterstock, Inc.

Screen shots on pages 27, 28, 29, 30, 32, 33, 34, 37, 39, 41, 42, 43, 46, 47, 48, 50, 51, 52, 56, 59, 60, 63, 65, 68, 69, 70, 72, 73, 74, 75, 77, 78, 79, 80, 81, 82, 84, 85, 86, 87, 88, 92, 93, 94, 95, 96, 100, 103, 104, 108, 109, 110, 111, 112, 115, 116, 118, 119, 121, 123, 124, 126 are reprinted courtesy of International Business Machines Corporation, © International Business Machines Corporation.

www.kendallhunt.com
Send all inquiries to:
4050 Westmark Drive
Dubuque, IA 52004-1840

Copyright © 2014 by Kendall Hunt Publishing Company

ISBN: 978-1-4652-4077-4

Printed in the United States of America
10 9 8 7 6 5 4 3 2 1

Content

INTRODUCTION

This book, for undergraduate students in political science, offers an introduction to applied statistics, which assists students in mastering one of the most widely used programs for statistical analysis (SPSS). Conventional wisdom suggests that navigating through SPSS is at times a challenging process usually accompanied by an anxiety associated with the array of statistical and mathematical concepts a student must confront. This book tackles "the fear of numbers" through examples and exercises whose aim is that of having the students apply these principles firsthand. Hence, its author does not neglect theories and definitions but rather integrates them pragmatically into the teaching of operations and procedures. In addition, this publication introduces SPSS by subjecting a variety of assumptions and situations to statistical analysis. The data set for this purpose is the 2012 General Social Survey (GSS). The GSS conducts basic scientific research on the American society's structure and development while monitoring societal change within the United States.

The main objectives of the book are (1) to help students gain a thorough understanding of applied statistics, (2) to provide a toolkit that allows students to conduct their own research, (3) to train students to explore and handle data in a systematic manner, and (4) to analyze these results with the help of SPSS.

After reading the book students should be able to:

- Describe and apply principles of research design and methodology, including systematic data gathering
- Apply all the statistical techniques presented in the book for the analysis of quantitative data
- Interpret, discuss, and present statistical information
- Understand and master the SPSS statistical software
- Carry out their own research and apply the necessary quantitative methods to address the given research question

PART I
DESCRIPTIVE STATISTICS

CHAPTER 1
Variables, Research Question, Hypothesis

The most important question stemming from the first lecture is about defining what a *variable* is. In simple words: a variable is any trait whose values change from case to case. For instance: gender, age, ethnicity, and sex are all variables that take different values in different situations. In other words, if we look at the variable *age*, the latter will vary from one student to the other.

Let us assume that we are interested in finding measures that would lower the levels of corruption within a country. Once we have identified our research interest, the next logical step would be to paraphrase the former as a *research question*. A research question (RQ) is a clear and focused question around which we will conduct our inquiry. Consequently, the RQ has to end with a question mark, and students should avoid phrasing it as a simple interrogative sentence.

Thus, the *research question* could be phrased in the following way:

- How can we decrease corruption in country X?

We can think of numerous factors that may contribute to decreasing corruption in country X. For instance, a more *efficient justice system* could lead to more convictions. The latter might in some instances increase the "costs" of corruption and undercut corruptive tendencies. Moreover, higher levels of *transparency* and *accountability* may equally be effective instruments to undermine corruption.

What is an "efficient justice system" or higher levels of "transparency" and "accountability" in this case? They are all *variables* that try to measure the changes in the level of corruption for a given country. Thus, *research questions* help students to focus their research by indicating a research path in addition to *narrowing the variables*, whose interaction may prove important for explaining suspected and uninvestigated trends.

The question then immediately follows: what type of variables are we narrowing down through a *research question*?

In statistics we distinguish between *dependent* and *independent* variables. The dependent variable in our example is the level of corruption in country X. The level of transparency and accountability, and the efficiency of the judiciary are all independent variables. Independent variables seek to explain the changes in our dependent variable. We can think of the independent variables as the "cause" versus the dependent as the "effect" or "result" in the given relationship. For instance, if country X improves on the efficiency of its judiciary in addition to holding state officials more accountable and increasing transparency, these changes will translate to lower levels of corruption.

In this case, we could say that *high levels of transparency and accountability in addition to an efficient justice system undermine corruption in country X.*

The statement that describes the relationship between the independent variables and the dependent is our **hypothesis**. The hypothesis takes the form of a specific statement of prediction, outlining our expectations about the relationship between the variables.

LEVELS OF MEASUREMENT

The next set of questions relate to the levels of measurement of the given variables. How can we quantify "transparency", an "efficient justice system", "accountability", and the "level of corruption"? What would the scores for these variables reflect? For instance, how could we understand a score of 5 for our "transparency" variable? Moreover, what kind of mathematical calculations are we allowed to carry out with them?

Thus, we distinguish between three levels of measurement:

- The Ordinal Level of Measurement
- The Interval - Ratio Level of Measurement
- The Nominal Level of Measurement

Ordinal Level of Measurement median and mode...

All four variables mentioned above (the three independents and the dependent variable) are measured at the *ordinal level*. Variables measured at the ordinal level have scores that can be rank-ordered in an ascending or descending fashion. For instance, we could measure accountability from a scale of 1 to 5, where 1 would stand for the lack of accountability in government decision making and 5 would indicate the highest level of accountability in the given country. A score of 3 would indicate a "more or less" accountable state administration. Thus, for each unit of analysis (in our case country), the variable *accountability* could encompass values from 1 to 5, allowing us to rank-order countries based on their level of accountability. Nonetheless, variables measured at the ordinal level face also serious limitations. For example, while we can rank-order the given countries on a scale from more to less accountable, we cannot precisely quantify the difference between a country that scored a 3 and a country that received a code of 1. We fail to measure the differences in the level of accountability between a country and another.

Another important variable measured at the ordinal level is socioeconomic status (SES), with three prominent categories: lower class, middle class, upper class. As with our "accountability" variable, we can rank-order the categories and claim that

[Note: SES/class is ordinal — subjective — easy to rank order but hard to understand diff btw a 3 and 5]

...iging to an upper class are considerably wealthier than those who ...er class (or middle class). However, we can't quantify the difference ...dual falling into one category (e.g., upper class) as opposed to another

...iables that capture respondents' level of satisfaction are measured ...el. For instance, a variable measuring the level of satisfaction with a given political party/regime would usually encompass three categories from "very satisfied", "satisfied", "neutral", "dissatisfied, to "very dissatisfied". While we can rank-order the given categories, we fail to attach a real value to where dissatisfaction ends and high satisfaction starts. Moreover, these categories are also extremely subjective. One respondent can indicate that he is very satisfied with the given political party, but at the end of the day, it would be hard to quantify the given subjective perception.

Interval-Ratio Level of Measurement

[Note: MATH → mostly only possible w/ Interval]

Variables measured at the interval-ratio level (such as age, number of children, income, height, weight) are very different from variabl... ...vel. Similarly to the latter, we can rank-order interval-ratio ... ...an rank-order students from the youngest to the oldest indi... ...In addition to rank-ordering age, we can also measure the difference between two scores (e.g., Respondent #5 is ten years older than Respondent #6). Behind each interval-ratio variable we have a real value (as opposed to just having a code) that allows us to carry out any desired mathematical calculation (e.g., calculating the Mean, Median, and the Mode).

[Note: mean, median, and mode]

A variable's level of measurement (nominal, ordinal, or interval-ratio) dictates the mathematical procedure to use. Since Hypothesis Testing I, II, and III contrast sample averages (e.g., sample average with population average, two or three sample averages) only interval-ratio variables (or in some instances correctly coded variables measured at the ordinal level) are allowed to be used for the above-mentioned Tests for Statistical Significance.

Nominal Level of Measurement

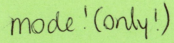

[Note: mode! (only!)]

In stark contrast to the interval-ratio level of measurement, variables measured at the nominal level (such as gender, area code, province of residence, religious affiliation, ethnicity, phone number, respondent number) *do not* allow us to carry out *any* mathematical calculations. The numbers that are displayed in our data set don't reveal "real" numbers, but the given variable's coding. For instance, the variable "gender" is usually coded as "1" for female and "2" for male (or vice versa). This code however does not represent a number for which we could calculate the mean or the median. Moreover, the code of "1" could be replaced by "22" or "555" or any other random number. Consequently, we cannot rank-order variables measured at the nominal level, nor carry out any mathematical calculations. The only calculation permissible with nominal variables is comparing the frequency of the categories (e.g., the most frequent value: the Mode). Thus, we could say that "females are more frequently represented in our sample than males; or the most frequent value is female".

CHAPTER 2

Measures of Central Tendency

Measures of central tendency "summarize" entire distributions of scores by describing the variable's most "common" value. The *Mode*, the *Median*, and the *Mean* are the three most well-known measures of central tendency.

THE MODE

The *Mode* is the value that occurs most frequently and is generally used with variables measured at the nominal level.

EXAMPLES

Let us look at the given distributions of scores:

1. *1, 2, 2, 3, 4*, and *5*: in the given distribution the most frequent value is *2*.
2. *Christian, Muslim, Orthodox*, and *Atheist*: in the given example we do not have a Mode. Often we can encounter distributions where we fail to report on the Mode.
3. *College, college, college, master's, master's, PhD, PhD, PhD*: the Mode is *college* and *PhD*. For some distributions we might have more than one Mode.
4. Let us look at the following distribution of grades: *A+, A+, A+, C, C, D, D-,* and *D-*. In this example, the most frequent value is *A+*. Does the mode reveal the central tendency of the given distribution? In other words, is this class an *A+* class on average? The correct answer is no. In this case, the Mode failed to capture the central tendency due to the distribution's positive skew.

Consequently, the Mode faces some serious limitations. We can encounter distributions of scores that do not have a Mode or have too many Modes, or in some situations the Mode fails to capture the distribution's central tendency.

THE MEDIAN

The *Median* always represents the distribution's center by dividing the former into two equal parts. Thus, half of the cases will have scores higher than the median and half of the scores will be lower than the Median.

For instance, if we report that the median income is $55,000/year, we automatically know that half of the respondents earn less than $55,000 and half of them earn more. Similarly, if the median age is 22 years then 50% of the scores will be lower (e.g., younger respondents) and the other half will be higher (e.g., older respondents). The Median is most frequently used when distributions are skewed and the Mode and the Mean fail to capture the distribution's central tendency.

EXAMPLE 1

When calculating the Median we need to first distinguish between an **even** or **odd** number of cases.

Let us assume that five random respondents (N=5) have disclosed their age. The given scores are illustrated below:

Respondent #1	Respondent #2	Respondents #3	Respondent #4	Respondent #5
22	21	17	18	25

In this example we have an *odd* number of cases. First, students will need to rank-order the given scores.

17	18	21	22	25

Mathematically, the Median's value is calculated by the formula: *(N+1)/2*. In our case, N=5, thus (5+1)/2=6/2=3. Consequently, the Median value will be the third value after we have rank-ordered the distribution's scores. Thus, the Median respondent in our distribution is 21 years old.

EXAMPLE 2

Let us introduce an additional respondent's age to the above illustrated distribution. The number of observations increases from five cases to six (N=6).

Respondent #1	Respondent #2	Respondent #3	Respondent #4	Respondent #5	Respondent #6
22	21	17	18	25	30

In this example we have an *even* number of cases. After rank-ordering the values we find that we have two middle values. The formula to find the first middle value for an even number of cases is *N/2*; thus 6/2 = 3; (the third value in our rank-ordered distribution). The second middle value is given by the formula: *N/2+1=6/2+1=3+1=4* (the fourth value in our rank-ordered distribution).

17	18	21	22	25	30

After we find the two middle values we have to calculate the average of the two. Thus, the Median respondent is (21+22)/2=43/2=21.5 years old.

Since the calculation of the Median requires us to rank-order the distribution's scores, we can calculate the latter for variables measured at the interval-ratio level and in some cases for variables measured at the ordinal level. *However, we are not allowed to calculate the Median for variables measured at the nominal level.*

EXAMPLE 3

We would like to calculate the Median for the ordinal variable socioeconomic status (SES). Our distribution encompasses five respondents who agreed to reveal their SES (N=5).

Respondent #1	Respondent #2	Respondent #3	Respondent #4	Respondent #5
Middle Class	Upper Class	Middle Class	Lower Class	Lower Class

After rank-ordering the above illustrated values, we find that the Median respondent belongs to the Middle Class.

Lower Class	Lower Class	Middle Class	Middle Class	Upper Class

EXAMPLE 4

If we decide to add an additional respondent's socioeconomic status (N=6), the distribution of scores will change in the following way:

Respondent #1	Respondent #2	Respondent #3	Respondent #4	Respondent #5	Respondent #6
Middle Class	Upper Class	Middle Class	Lower Class	Lower Class	Upper Class

We rank-order the above illustrated cases and find that the Median respondent belongs to the Middle Class. However, the only reason why we could calculate the Median in the first place is because both median values located at the distribution's center share a similar SES.

Lower Class	Lower Class	Middle Class	Middle Class	Upper Class	Upper Class

EXAMPLE 5

The calculations would change considerably if the additional respondent would display a different SES (e.g., Lower Class).

Respondent #1	Respondent #2	Respondent #3	Respondent #4	Respondent #5	Respondent #6
Middle Class	Upper Class	Middle Class	Lower Class	Lower Class	Lower Class

Due to the divergent two middle values, we find that we are unable to calculate the above illustrated distribution's Median score (e.g., the respondents have a different SES).

Lower Class	Lower Class	Lower Class	Middle Class	Middle Class	Upper Class

Consequently, with variables measured at the ordinal level that encompass an even number of observations, we are able to calculate the Median solely if the two middle values belong to the same category.

THE MEAN

The third and most prominent measure of central tendency is the *Mean*, or the arithmetic average. Calculating the Mean is quite simple: students have to add up all the scores and then divide the former by the total number of cases. The Mean should be calculated only for variables measured at the interval-ratio level and, at times, for correctly coded ordinal variables.

$$\bar{X} = \frac{\sum_{i=1}^{n} X_i}{n}$$

EXAMPLE

Let us calculate the Mean for the following distribution of scores: *1, 6, 3, 6,* and *2*.

$\bar{x}=(1+6+3+6+2)/5=18/5=3.6$

CHAPTER 3
The Normal Distribution

In statistics, we often assume that variables measured at the interval-ratio level are normally distributed. The latter assumption is essential for carrying out Tests for Statistical Significance, discussed in the second part of the book.

GRAPH 1

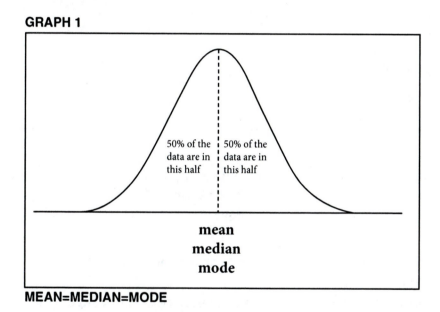

50% of the data are in this half | 50% of the data are in this half

mean
median
mode

MEAN=MEDIAN=MODE

Graph 1 displays the distribution of a society's IQ level. We immediately observe the bell shape of the curve encompassing "tails" that are spreading infinitely in both directions. The tails include some individuals (or outliers) with extreme scores for both measures. That is, we would find few individuals in a society who have either a very low IQ (i.e., less than 70) or an extremely high IQ (i.e., around 150).

The bell shape of the curve illustrates that the highest frequency is clustered around average values. In other words, the bulk of the people will have an average IQ (i.e., from 90 to 110) with frequency decreasing as we approach either extreme.

Similar assumptions of normal distribution hold for height, weight, and grades (to just name a few). The Mean, Median, and Mode of normally distributed interval-ratio variables encompass either the same values or values that are very close to each other.

If the Median is higher or lower than the Mean and the Mode, the distribution has either a negative or positive skew. For instance, income is positively skewed, or age in ageing societies (or weight in obese societies) is negatively skewed.

POSITIVELY SKEWED DISTRIBUTIONS

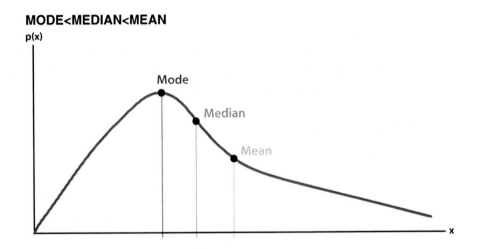

In many countries income is positively skewed. A positive skew in income distribution reveals a significant gap between the poor and the rich. The Gini coefficient, commonly used as the measure of income/wealth inequality, will increase in such instances[1]. Inevitably, we observe that:

- The Mean is higher than the Median and higher than the Mode (Mean>Median>Mode).
- The long tail that extends to the right exposes extremely high values (e.g., millionaires, billionaires) whose colossal wealth artificially inflates the Mean.
- The Mode (the most frequent value) comprises people living in poverty or very close to the poverty threshold. As we would expect, the number of poor people in such countries is very high.

As a result, the correct measure of central tendency is no longer the Mean but the Median (the measure that divides the distribution into two equal halves). Reporting on the Mean (e.g., per capita GDP or average income) would give a false picture of where the "average" citizen may be located along the income distribution curve. Positively

---

1   For more studies on income inequality and how to measure poverty, see also http://web.worldbank. org/WBSITE/EXTERNAL/TOPICS/EXTPOVERTY/EXTPA/0,,contentMDK:22405907~menuPK:6626650~ pagePK:148956~piPK:216618~theSitePK:430367,00.html

skewed distributions can be *normalized* by eliminating outliers from the data set (e.g., deleting extremely high or extremely low scores).

EXAMPLE

Let us assume we ask nine individuals living in the United States to disclose their income. The distribution of scores would take the following form:

Respondent #1: unemployed	Respondent #6: $58,000
Respondent #2: $10,000	Respondent #7: $65,000
Respondent #3: $10,000	Respondent #8: $67,000
Respondent #4: $50,000	Respondent #9: $1,000,000
Respondent #5: $55,000	

The Mode

The Mode in our distribution is $10,000 (Respondent #2 and Respondent #3's income is the most frequent value), while the Median is $55,000 (see calculations below).

The Median

- First, rank-order the values.

Respondent Number	#1	#2	#3	#4	#5	#6	#7	#8	#9
Value	0	10,000	10,000	50,000	55,000	58,000	65,000	67,000	1,000,000

- Second, decide if you have an *even* or *odd* number of cases.

The total number of respondents is nine, thus we have an *odd* number of cases. Consequently, the Median value will be Respondent #5's income, which is $55,000.

The Mean

In contrast to the Median, when calculating the Mean, we find that the distribution's "average" income is much higher than the Mode and higher than the Median.

$$\bar{x} = (0 + 10,000 + 10,000 + 50,000 + 55,000 + 58,000 + 65,000 + 67,000 + 1,000,000)/9 = \$146,111.11$$

But does a sample average of approximately $146,111 represent the central tendency of the above illustrated distribution? Unfortunately, it does not. The Mean income is artificially *inflated* by the colossal wealth of Respondent #9. The latter is earning $1 million/year while other respondents in our distribution are struggling financially. In this case, the correct measure of central tendency is the Median ($55,000).

NEGATIVELY SKEWED DISTRIBUTIONS

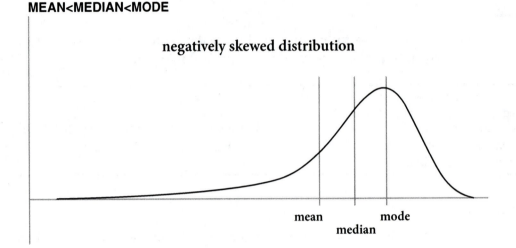

MEAN<MEDIAN<MODE

With distributions that are negatively skewed, the situation is reversed. For instance, in ageing societies (countries with a significant number of elderly people), age distribution is negatively skewed, with a tail extending to the left. Consequently, we observe that:

- The Mean is lower than the Median and lower than the Mode (Mean<Median<Mode).
- The long tail that extends to the left exhibits extremely low values (e.g., newborns, toddlers, children, teenagers) whose comparatively young age artificially deflates the mean.
- The Mode (the most frequent value) comprises elderly individuals. The latter is extremely high in countries with a predominantly ageing population.

Similar to positively skewed distributions, the correct measure of central tendency is the Median. In contrast to the middle value, the Mean age indicates a much younger "average" citizen than the actual central tendency. The Mean age in this case is lower than the Median and the Mode, as a result of the long tail extending to the left and encompassing much younger individuals than the most frequent value.

EXAMPLE

Let us assume we ask ten random individuals living in Germany to disclose their age. The distribution of scores would take the following shape:

Respondent #1: infant	Respondent #6: 50 years old
Respondent #2: 2 years old	Respondent #7: 60 years old
Respondent #3: 3 years old	Respondent #8: 80 years old
Respondent #4: 4 years old	Respondent #9: 80 years old
Respondent #5: 40 years old	Respondent #10: 85 years old

The Mode

The Mode in our distribution is 80 (Respondent #8 and Respondent #9), while the Median is 45 years old (the average value of the two middle values, Respondents #5 and #6; see calculations below).

The Median

- First, rank-order the values: infant (0 years), 2, 3, 4, 40, 50, 60, 80, 80, 85

#1	#2	#3	#4	#5	#6	#7	#8	#9	#10
0	2	3	4	40	50	60	80	80	85

- Second, decide if you have an *even* or *odd* number of cases.

We have asked ten respondents, thus we have an *even* number of cases. Consequently, the two Median values will be Respondent #5's age and Respondent #6's age (Respondent #5=N/2= 10/2; Respondent #6=N/2 +1: #5+1=#6)

- Third, calculate the average for the two Median values.

Median = (40+50)/2= 90/2= 45 years old

The Mean

In contrast to the Median, when we calculate the Mean, we find that the distribution's "average" age is much lower than the Mode and lower than the Median.

$$\bar{x} = (0+2+3+4+40+50+60+80+80+85)/10 = 40.4 \text{ years}$$

But does an average age of 40.4 years represent the central tendency of the distribution? The correct answer in this case is no. The former is artificially *deflated* by very low values (the ages of Respondents #1, #2, #3, and #4) in our sample. Consequently, the correct measure of central tendency is the Median (45 years).

CONCLUSION

The Mean always moves toward the extreme scores. If the outliers include extremely high scores, the Mean artificially inflates. In the reverse situation, when the outliers comprise extremely low scores, the Mean artificially deflates (or decreases). In order to know the direction and distribution of a skew, we need to compare the Mean and the Median.

IF THE:

MEAN>MEDIAN → POSITIVE SKEW
MEAN<MEDIAN → NEGATIVE SKEW

CHAPTER 4
Measures of Dispersion

THE STANDARD DEVIATION

Let us assume we would like to go out for a movie or take our friend out to a great restaurant. After looking up some ratings on IMDB.com, we narrow down the choice to two movies. Movie A has an average rating of 6.78/10, and Movie B has an average rating of 6.67/10. Which movie are you more likely to see?

Similarly, we Google the best restaurants on Yelp.com and come up with two choices: Restaurant A had an average rating of 4.5/5, and Restaurant B had an average rating of 4.7/5. Do we choose restaurant A or B? Alternatively, let's say that before traveling with our parents, we are searching for a very good hotel. After we browsed through Tripadvisor.com, we have singled out two hotels: one's rating is slightly above the other's: out of 10 stars the first hotel received 8, and the second hotel had an average rating of 8.1/10. Where will we check-in, and where will we dine?

The aforementioned questions arise on a daily basis without us knowing that the two measures we rely on mostly are *the Mean* (measure of central tendency) and *the Standard Deviation* (measure of dispersion). While measures of central tendency "summarize" our distribution and present us with the central score, often knowing the average does not provide us with enough information to take the best decision. In these instances, calculating (or observing) a measure of dispersion becomes extremely important.

EXAMPLE

Let us assume that nine people have rated Movie A and Movie B on IMDB.com. Movie A's average rating is 6.78/10 while Movie B's average is 6.67/10. Based on the measure of central tendency, we should go and watch Movie A because of its higher rating (6.78>6.67). But is the average rating the only thing we care about when comparing the two movies? Not if we want to take the best decision possible.

Let us give a closer look at the given ratings:

MOVIE A								
1st Rating	2nd Rating	3rd Rating	4th Rating	5th Rating	6th Rating	7th Rating	8th Rating	9th Rating
10	8	1	1	9	10	10	10	2

MOVIE B								
1st Rating	2nd Rating	3rd Rating	4th Rating	5th Rating	6th Rating	7th Rating	8th Rating	9th Rating
7	7	6	6	7	7	8	6	6

While both movies have received a close to equal average rating (6.78 for Movie A and 6.67 for Movie B), the nine respondents strongly disagreed while rating Movie A and strongly agreed while rating Movie B. Movie A's ratings include scores as low as 1 and 2 (indicating that Respondents #3, #4, and #9 did not enjoy the movie at all) or as high as 10 and 9 (indicating that Respondents #1, #5, #7, and #8 had a high opinion of the motion picture). If we contrast these results with Movie B's ratings, we find that the latter has fewer outlier scores (respondents who loved or hated the movie) and the majority of the ratings cluster around the average. In other words, people agreed that it was a good movie, giving it scores from 6, 7, to 8. If you are risk averse, you definitely opt for Movie B because there is a greater likelihood that your preferences will align with the given ratings you can see in the table illustrated above (Movie B).

Choosing Movie A, which has a slightly higher average rating (6.78-6.67=0.11) would also entail a significant amount of risk, as you might agree with those who disliked it and gave it a very low rating or those who believed it is one of the best movie they have ever watched. In this case, the smarter choice would be to go and watch the movie that has a lower standard deviation (Movie B).

Measures of dispersion indicate the extent of the scores' dispersion around the mean. Distributions with low standard deviations cluster around the average, while distributions with high standard deviations have scores that encompass the whole range of possible scores.

GRAPH 1

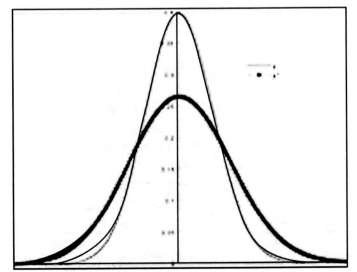

Graph 1 displays both movies' ratings and averages. While the average ratings are extremely close, the peakness of the distributions is very different. In other words, while respondents agreed that Movie B was a good movie (thinner line with the higher peak), the ratings for Movie A were spread from the lowest to the highest distribution (thicker black line). The curve for Movie A is lower and steeper because the distribution encompasses values revealing dissatisfaction (ratings of 1 and 2), positive reception (8), and enthusiastic appreciation (ratings of 9 and 10). Consequently, *the standard deviation increases in value as the distribution becomes more variable and decreases in value as the distribution is more similar* (e.g., respondents "agree").

Let us calculate the standard deviation for both distributions (Movie A and Movie B).

The Formula

$$\sigma = \sqrt{\frac{\sum \left(x - \overline{x} \right)^2}{N}}$$

The standard deviation for Movie A is 3.91 while the standard deviation for Movie B is 0.66. A higher standard deviation for the former reveals a higher dispersion among the scores (e.g., greater "disagreement") while a lower standard deviation for Movie B is attributable to the ratings' closeness (e.g., greater agreement among raters).

TABLE 1

The Scores X_i	Deviations $x_i - x$	Deviations Squared $(x_i - \overline{x})^2$	Sum of Deviations Squared $\sum (x_i - \overline{x})^2$
10	(10-6.78) = 3.22	10.37	10.37
8	(8-6.78) = 1.22	1.49	1.49
1	(1-6.78) = −5.78	33.41	33.41
1	(1-6.78) = −5.78	33.41	33.41
9	(9-6.78) = 2.22	4.93	4.93
10	(10-6.78) = 3.22	10.37	10.37
10	(10-6.78) = 3.22	10.37	10.37
10	(10-6.78) = 3.22	10.37	10.37
2	(2-6.78) = −4.78	22.85	22.85
		TOTAL	137.56/N=15.28

Standard Deviation for Movie A = $\sqrt{15.28}$ = 3.91

TABLE 2

The Scores x_i	Deviations $x_i-\bar{x}$	Deviations Squared $(x_i-\bar{x})^2$	Sum of Deviations Squared $\Sigma (x_i-\bar{x})^2$
7	(7-6.67) = 0.33	0.11	0.11
7	(7-6.67) = 0.33	0.11	0.11
6	(6-6.67) = −0.67	0.45	0.45
6	(6-6.67) = −0.67	0.45	0.45
7	(7-6.67) = 0.33	0.11	0.11
7	(7-6.67) = 0.33	0.11	0.11
8	(8-6.67) = 1.33	1.77	1.77
6	(6-6.67) = −0.67	0.45	0.45
6	(6-6.67) = −0.67	0.45	0.45
		TOTAL	4/N=4/9=0.44

Standard Deviation for Movie B = $\sqrt{0.44}$ = 0.66

Three important questions follow:

- What do deviations measure mathematically?
- Why do we have negative deviations?
- Why do we need to square the deviations?

What do deviations measure mathematically?

An integral part of the standard deviation formula is the calculation of the "Deviations" or distances from the mean by subtracting the mean from each distribution's score (see the second column of Table 1 and 2). Mathematically, the formula $X_i-\bar{x}$ illustrates each score's distance from the mean. The further the scores fall from the average (e.g., the higher the "disagreement" among the ratings), the larger such distances will be.

Why do we have negative deviations?

When we subtract the average from each score, some deviations (or distances from the mean) will result in a negative score as a result of scores that are lower than the average. For any distribution, we will have scores that will be higher than the average (above-average scores) and vice versa (below-average scores). When we subtract from a lower score a higher average, the result will be negative. For instance, when calculating the standard deviation for Movie B, some ratings are lower than the average rating. (e.g.,

In row #3 we have to subtract from a rating of 6 the mean score of 6.67, resulting in a negative score of -0.67.)

Why do we need to square the deviations?

The explanations outlined above contain the answer to this question. While the deviations measure the distances from the average, such distances can attain negative values due to lower-than-average scores. In order to circumvent this problem (e.g., remove the negative sign), we will first need to square the deviations.

PART II

INFERENTIAL STATISTICS

CHAPTER 5

Hypothesis Testing I: The One-Sample Case

Chapter 5 will begin with a general overview of *Significance Testing* and present three SPSS exercises introducing concepts such as *statistical significance*, *Null Hypothesis*, *Research Hypothesis*, and *random chance*. In this chapter we focus on the *One-Sample Case*, comparing a random sample against a population.

In the first exercise, we are interested to see if the income of Americans who have indicated an exciting lifestyle is *significantly different* (e.g., higher) than the income of the overall American population. In Exercise 2, we will compare the income of a sample of non-U.S. citizens residing in the United States to the average income of all Americans. Numerous studies expose various obstacles and challenges immigrants face on the U.S. labor market. Such findings prompt us to inquire whether non-U.S. citizens' earnings are *significantly different* (e.g., lower) vis-à-vis those of the U.S. population as a whole. Are non-U.S. citizens facing serious obstacles and challenges in the American labor market?

The third exercise introduces the *recoding* procedure, followed by a *One-Sample T Test*. The test's purpose is that of comparing the number of children of respondents who have indicated to be dissatisfied with their family life to the mean number of an average American family's children. Would individuals who expressed their dissatisfaction with their family life also have fewer children than the average American family?

CONDUCTING A ONE-SAMPLE TEST IN SPSS

Exercise 1

The three exercises and illustrations below use the 2012 General Social Survey (GSS) data set[1].

---

1 http://www.norc.org/Research/Projects/Pages/general-social-survey.aspx

You can download the data from http://www3.norc.org/GSS+Website/Download/ SPSS+Format/ .

Significance testing, or *hypothesis testing*, reveals the probability that a relationship between two variables reflects observed trends in the population. Otherwise the relationship between the two variables was caused by random chance alone. The goal of the researcher is to reject the Null Hypothesis, a statement that directly contradicts the Research Hypothesis. The latter is conveying the researcher's assumption, and his/her goal is to strengthen its evidence in support of the Research Hypothesis. The first step toward this goal is to reject the Null Hypothesis of "no difference". By rejecting the Null Hypothesis, the researcher has proven that the difference between the sample mean and the population mean is statistically significant and has not been caused by random chance alone.

In the first exercise, we will carry out a One-Sample Test in order to compare the income of a sample of Americans who have indicated that their life is "exciting" to the average income of all Americans. We would assume that individuals who live a happy/exciting life will have a higher income than the "average" American.

The Research Hypothesis (H$_1$)

The researcher would like to demonstrate that the mean sample income of respondents who live an exciting lifestyle is significantly different (e.g., higher) than the average income of the overall American population.

The Null Hypothesis (H$_0$)

The Null Hypothesis states that the mean sample income of respondents who live an exciting lifestyle is "no different" than the average income of the overall American population.

If we are able to reject the Null Hypothesis, we can conclude that the mean income for Americans living "exciting" lives is significantly different (e.g., higher) than that of the general population.

First, let us look at the ordinal variable that captures respondents' life satisfaction. Respondents were asked a closed-ended question and were presented with three possible answers: (1) Exciting, (2) Routine, and (3) Dull. Due to the missing values, three additional categories were created: (0) Inapplicable, (8) Do not know, and (9) Not applicable. If we click on VALUES the following VALUE LABELS window will open, displaying the variable's categories.

VALUE LABELS

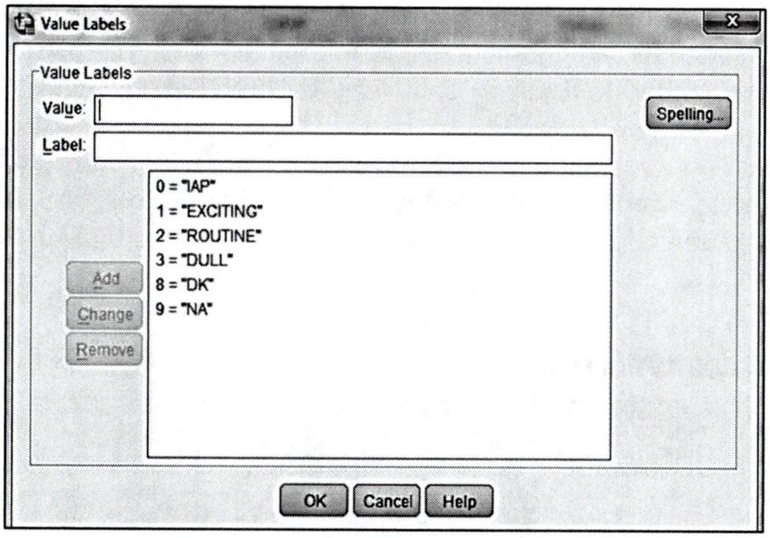

Handwritten margin notes:

DATA
↓
Select Cases
↓
Variable and Category
(select, arrow, type code)
↓
Continue
+
OK

In order to select the sample of respondents who have indicated to live "Exciting" lives, we need to look up the exact code for the given category and carry out an IF CONDITION.

To carry out an IF CONDITION, we use the SELECT CASES command under the DATA menu bar option. The SELECT CASES IF window will allow the researcher to select the variable and the given category (e.g., sample subset) he/she is interested to study. In our case, the variable's name is "life" and the sample subset distinguishing between respondents who have indicated to live "exciting" lives has received a code of "1".

Thus, we select the given variable, click on the arrow to move "life" into the text box and type "=1". A code of "1" for the variable "life" encompasses the subset of respondents who are living exciting lives. Click on CONTINUE (the SELECT CASES IF window will close automatically) and click on OK (the SELECT CASES window will close as well).

SELECT CASES: IF CONDITION IS SATISFIED

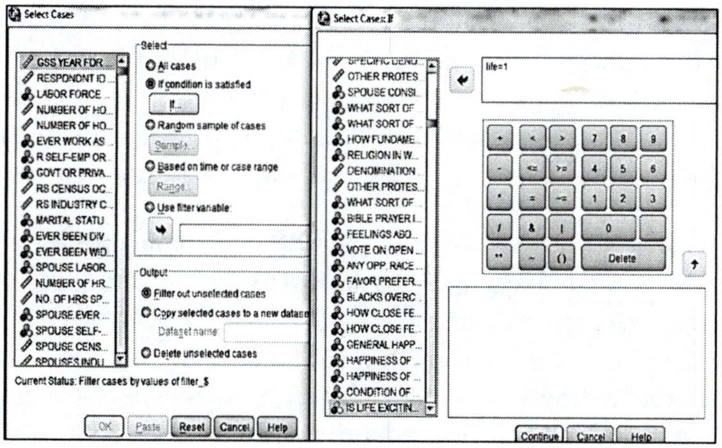

If we click on DATA VIEW after we have carried out the SELECT CASES command, we can see that respondents who do not have "Exciting" lifestyles will be temporarily eliminated from our data set. For instance, Respondents #4–9, 11–13, 15–20, 23, and so forth have either pointed to living a "Dull" or "Routine" existence or failed to answer the question.

It is important to note that the IF CONDITION is a "sticky condition"; that is, after students have carried out Hypothesis Testing I, they need to return to the SELECT CASES command and click on ALL CASES in order to include all the cases in the forth-coming analysis.

CASES EXCLUDED FROM OUR SAMPLE

File   Edit   View   Data   Transform   Analyze   Direct Marketing   Graphs   Utilities   Add-ons   Window   Help

	year	id	wrkstat	hrs1	hrs2	evwork	wrkslf	wrkgovt	
1	2012	1	2	15	-1	0	2	2	
2	2012	2	2	30	-1	0	2	2	
3	2012	3	1	60	-1	0	2	2	
4	2012	4	8	-1	-1	1	2	2	
5	2012	5	5	-1	-1	1	2	1	
6	2012	6	8	-1	-1	1	2	2	
7	2012	7	7	-1	-1	1	2	2	
8	2012	8	7	-1	-1	2	0	0	
9	2012	9	7	-1	-1	2	0	0	
10	2012	10	1	40	-1	0	2	2	
11	2012	11	8	-1	-1	1	2	2	
12	2012	12	2	20	-1	0	8	2	
13	2012	13	7	-1	-1	2	0	0	
14	2012	14	8	-1	-1	1	2	1	
15	2012	15	2	32	-1	0	2	2	
16	2012	16	1	53	-1	0	2	2	
17	2012	17	1	60	-1	0	2	2	
18	2012	18	1	40	-1	0	2	2	
19	2012	19	1	40	-1	0	2	2	
20	2012	20	6	-1	-1	2	0	0	
21	2012	21	4	-1	-1	0	2	1	
22	2012	22	1	12	-1	0	2	2	
23	2012	23	1	40	-1	0	2	1	

Before carrying out the One-Sample T Test, students should acquaint themselves with the variable that captures respondents' income. While income is usually measured at the interval-ratio level, in the 2012 GSS data set, "Respondents' Income" is coded as an ordinal variable. Therefore, variable "incom06" has 26 categories (in addition to the categories "don't know" coded as "98", "not applicable" coded as "99", and "0" as "inapplicable").

By clicking on VALUE under VARIABLE VIEW, we can observe the categories of "incom06". Thus, respondents who have indicated to earn less than $1,000 have received a code of "1". Similarly, respondents earning from $40,000 to $49,990 are included in the category coded as "18", and those making $150,000 or more were coded as "25".

If, for instance, the coding would fail to follow a logical ascending trend as income increases (e.g., those making $150,000 or more would have received a code of "5" while

Need to recode if not in proper order

those earning $40,000 to $49,990 would have received a code of "12", and those having no or little income a code of "10"), we would have to recode the variable in order to align the increasing or decreasing tendencies to the increasing/decreasing order of the codes attributed to each category.

VALUE LABELS

Value Labels

Value:

Label:

Spelling...

```
0 = "IAP"
1 = "UNDER $1 000"
2 = "$1 000 TO 2 999"
3 = "$3 000 TO 3 999"
4 = "$4 000 TO 4 999"
5 = "$5 000 TO 5 999"
6 = "$6 000 TO 6 999"
7 = "$7 000 TO 7 999"
8 = "$8 000 TO 9 999"
9 = "$10000 TO 12499"
10 = "$12500 TO 14999"
11 = "$15000 TO 17499"
12 = "$17500 TO 19999"
13 = "$20000 TO 22499"
14 = "$22500 TO 24999"
15 = "$25000 TO 29999"
16 = "$30000 TO 34999"
17 = "$35000 TO 39999"
18 = "$40000 TO 49999"
19 = "$50000 TO 59999"
20 = "$60000 TO 74999"
21 = "$75000 TO 89999"
22 = "$90000 TO 109999"
23 = "$110000 TO $129999"
24 = "$130000 TO $149999"
25 = "$150000 OR OVER"
26 = "REFUSED"
```

Add   Change   Remove

OK   Cancel   Help

[handwritten margin notes:] Analyze ↓ compare Means ↓ One Sample T Test — then set test variable ↓ move to test variables using arrow

Following the sample selection, we will carry out the One-Sample T Test procedure. In order to test if the mean income of respondents living an exciting lifestyle differs from the average income of all Americans, we perform a One-Sample T Test by selecting ANALYZE -> COMPARE MEANS -> ONE-SAMPLE T TEST from the main menu bar.

Subsequently, the ONE-SAMPLE T TEST dialogue box will open, allowing us to select the given interval-ratio (or correctly coded ordinal variable) as the TEST VARI-ABLE. The T Test will contrast the average income of our sample of Americans living an exciting lifestyle with the mean income of all Americans. Thus, we select the variable "incom06" (Respondents' Income) and move the given variable to our TEST VARIABLE(S) box by clicking on the arrow located between the variable list and the TEST VARIABLE(S) box.

ONE-SAMPLE T TEST

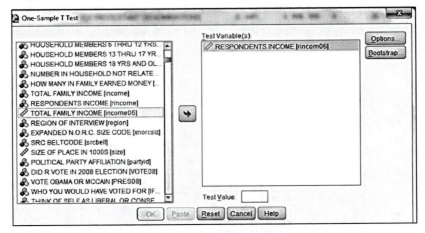

In addition, we also observe the TEST VALUE box requiring the researcher to enter the interval-ratio or ordinal value that best approximates the average income of "all" Americans. Thus, we will have to look up the mean or median population income for the year of 2012.

According to recent statistics, the American median annual household income is $52,100.[2] It is important for students to remember that because our test variable (Respondent's Income, variable "income06") is measured at the ordinal level, the above-mentioned median income has to be expressed as an ordinal category as well. Consequently, we have to look up the category that encompasses the $52,100 median income value. According to the VALUE LABELS window, the latter will fall into an income category that received a code of "19". Thus, the researcher will enter this code into the TEST VALUE box.

ONE-SAMPLE T TEST: TEST VALUE ENTERED

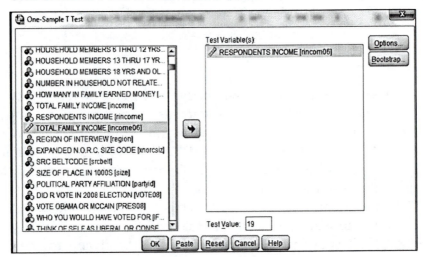

---

2   http://www.nytimes.com/2013/08/22/us/politics/us-median-income-rises-but-is-still-6-below-its-2007-peak.html?_r=0

Following the selection of the test variable and the test value, SPSS will generate two output tables (Table 1 and 2 illustrated below).

The ONE-SAMPLE STATISTICS table presents us with some descriptive statistics, such as the number of respondents who have indicated to live an exciting life (N=923) and their average income (15.38). The latter falls somewhere between the categories coded as "15" (containing incomes from $25,000 to $29,999) and "16" (comprising incomes from $30,000 to $34,999).

If we compare the sample mean illustrated in Table 1 with the population mean ($52,100 expressed as an ordinal category), we observe a mathematical difference of approximately $25,000 on average. The question then logically follows: *Is this difference between the sample mean and the population mean statistically significant, or is it caused by random chance alone?*

In order to answer this question we need to interpret the results illustrated in the second output table (Table 2: ONE-SAMPLE TEST).

Table 2 displays a T (obtained) score of -17.887 and a "Sig. (2 –tailed)" value of 0.000. The significance is the exact probability that the observed mathematical difference (of approximately $25,000) between the sample mean and the population average was caused by random chance alone. In other words, the Sig. value supports our Null Hypothesis—that is, the statement that the population of Americans living an exciting lifestyle is "no different" from the American population as a whole.

TABLE 1

One-Sample Statistics				
	N	Mean	Std. Deviation	Std. Error Mean
Respondents Income	923	15.38	6.146	.202

TABLE 2

One-Sample Test						
	Test Value=19					
					95% Confidence Interval of the Difference	
	t	df	Sig. (2-tailed)	Mean Difference	Lower	Upper
Respondents Income	−17.887	922	.000	−3.619	−4.02	−3.22

If the significance is lower than 0.05 (assuming an alpha of 5% and a confidence level of 95%), we can reject the Null Hypothesis of "no difference". On the other hand, if the Sig. value is higher than 0.05, we fail to reject the Null Hypothesis.

The significance outlined in Table 2 (ONE-SAMPLE TEST OUTPUT) is 0.000. The latter is much lower than 0.05, allowing us to reject the Null Hypothesis. Consequently, we can reject the statement of "no difference" and conclude that the difference in the population average (approximated by the median income score) and the mean sample income is significant. There is a significant difference between the population of Americans living an exciting lifestyle in contrast to the population of Americans as a whole.

Exercise 2

In the following exercise, we will carry out a One-Sample Test in order to compare the income of a sample of non-U.S. citizens living in the United States to the average income of all Americans.

Let us first examine the nominal variable that captures respondents' immigration status. If we click on the VALUES option for the variable "USCITZN", the following VALUE LABELS window will open.

VALUE LABELS

The VALUE LABELS window displays the nominal variable's seven categories ("USCITZN"). Respondents could indicate if they were "U.S. Citizens", coded as a "1"; "Not U.S. Citizens", receiving a code of "2"; or codes "3" and "4", if he or she declares to be a U.S. citizens born in Puerto Rico, the U.S. Virgin Islands, or outside the United

States to parents who were U.S. citizens. The categories (0) Inapplicable, (8) Don't Know, and (9) Not Applicable were also included.

Numerous studies expose various obstacles and challenges immigrants face on the U.S. labor market. Such findings prompt us to inquire whether non-U.S citizens' earnings are significantly different (e.g., lower) vis-à-vis those of the U.S. population as a whole. The Null Hypothesis (H_0) states that the average income of all non-U.S. citizens is "no different" from the national average. In contrast to the H_0, the Research Hypothesis articulates that the earnings of non-U.S. citizens are "different from" the American average income.

In order to reject the Null Hypothesis, we have to carry out a One-Sample T Test. Similarly to Exercise 1, first we select a sample of non-U.S. citizens with the help of an IF CONDITION.

Clicking on DATA -> SELECT CASES from the menu bar will open the SELECT CASES window. The researcher has the opportunity to select the nominal variable "USCITZN" (Respondents' Immigration Status) and specify that solely respondents who have received a code of "2" should be included in our sample. (USCITZN=2, indicating that they are not U.S. citizens).

SELECT CASES: IF CONDITION IS SATISFIED

Data
↓
Select
Cases
↓
USCITZN
(specify
that
we want
code 2)
↓
Continue
↓
OK

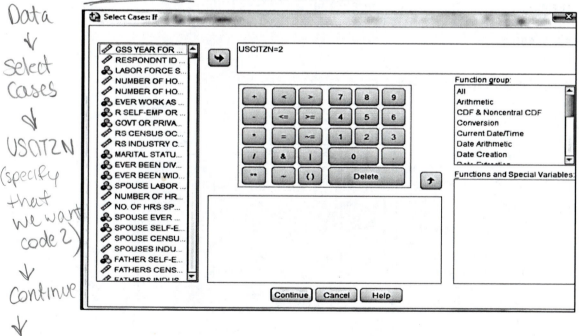

After clicking on CONTINUE and OK, both SELECT CASES: IF and the SELECT CASES window will close. Returning to the DATA VIEW, the researcher can observe the cases that were excluded from the sample.

EXCLUDED CASES

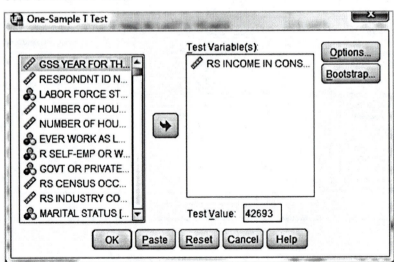

Following the selection of our sample that includes all non-U.S. citizens, we will carry out a One-Sample T Test by clicking on ANALYZE -> COMPARE MEANS-> ONE-SAMPLE T TEST.

ONE-SAMPLE T TEST

The researcher will now move the interval-ratio variable into the TEST VARIABLE(S) box in order to compare the sample mean with the average population value. In Exercise 1, we have selected variable "rincom06" (Respondents' Income), an interval-ratio variable recoded as an ordinal with 20+ categories. In this exercise, we will replace variable "rincom06" (Respondents' Income) with variable "realrinc" (Respondent's Income in Constant Dollars). The latter is the original interval-ratio variable that measured respondents' income in absolute terms, foregoing the use of categories. Consequently, when entering the TEST VALUE—that is, the value representing the average income of the American population as a whole—we will use the 2012 per capita personal income[3]. According to the U.S. Department of Commerce, in 2012, the average American earned $42,693 a year. After entering the given value in the TEST VALUE box and clicking on OK, two output tables will be generated (Table 3 and 4 illustrated below).

The ONE-SAMPLE STATISTICS table (Table 3) provides us with some basic information about our sample of 162 non-U.S. citizens, who earn on average $19,473. If we contrast this average annual income to the overall American average, we observe that non-U.S. citizens earned approximately $23,220 less than the "average" American.

The questions then logically follow: Is this difference *statistically significant*, or is it attributable to *random chance* alone? Can we generalize from our 162 observations and extend the research hypothesis to the whole population of non-U.S. citizens? Indeed, does the population of non-U.S. citizens earn significantly less than the average

"Is the difference STATISTICALLY SIGNIFICANT or is it attributable to random chance alone?"

TABLE 3

One-Sample Statistics				
	N	Mean	Std. Deviation	Std. Error Mean
Rs Income in Constant $	162	19473.21	38998.380	3064.002

TABLE 4

One-Sample Test						
	Test Value=42693					
					95% Confidence Interval of the Difference	
	t	df	Sig. (2-tailed)	Mean Difference	Lower	Upper
Rs Income in Constant $	−7.578	161	.000	−23219.792	−29270.61	−17168.98

---

3 Per capita personal income is total personal income divided by total midyear population. Source: U.S. Dept. of Commerce, Bureau of Economic Analysis. Released March 2013. Available from: http://bber. unm.edu/econ/us-pci.htm

American? Or put differently: Is the average income of all non-U.S. citizens in America "no different" than the national average?

In order to answer these questions and be able to take an informed decision about rejecting or accepting the Null Hypothesis, we have to look at the output in our ONE-SAMPLE TEST table. The researcher would like to reject the Null Hypothesis, a statement that directly contradicts the Research Hypothesis. The latter is conveying the researcher's assumption, that non-U.S. citizens on average have lower incomes and are from a poorer socioeconomic background in contrast to the national average. At this point, the researcher's goal is to strengthen the evidence in support of the Research Hypothesis. The first step toward this goal is to reject the Null Hypothesis of "no difference". By rejecting the Null Hypothesis, the researcher has proven that the difference between the mean income of non-U.S. citizens and the national average is statistically significant and has not been caused by random chance.

In order to answer the above mentioned questions, we will have to look up the "Sig. (2-tailed)" value. The latter reflects the probability of the relationship between the variable "realrinc" (Respondents' Income) and the variable "USCITZN" (Respondents' Citizenship) to have been caused by random chance alone. The "Sig. (2-tailed)" value is the probability of getting the observed difference of $23,220 between the sample mean of $19,473 and the population mean of $42,693, if only random chance would be operating. Or in other words, the significance reflects the probability that the Null Hypothesis of "no difference"—that is, the statement that there is "no difference" between the sample mean and the population mean—would be correct.

If the "Sig. (2-tailed)" value is lower than our alpha level of 5%, we are allowed to reject the Null Hypothesis. Conversely, if the significance will be higher than 0.05, we fail to reject the Null Hypothesis, which posits that the average income of the non-U.S. citizens' population is "no different" than the average income of the entire U.S. population.

In our case the "Sig. (2-tailed)" value is 0.000, lower than 0.05 (0.000<0.05), thus we can reject the Null Hypothesis. Consequently, the observed difference of $23,220 cannot be attributed to random chance and the given difference is statistically significant. Thus, we can generalize from 162 observations in our sample to the population of non-U.S. citizens and claim that unfortunately the former still faces many obstacles and challenges in the American labor market.

Exercise 3

In this last exercise we will introduce the recoding procedure, followed by a One-Sample Test to contrast respondents' number of children who have indicated to be dissatisfied with their family life with the mean number of children of an average American family in 2012. Would individuals who expressed their dissatisfaction with their family life also tend to have fewer children than the average American family?

The VALUE LABELS window below illustrates the ordinal variable's ("satfam7") coding capturing respondents' level of family satisfaction. Thus, respondents could indicate to be "Completely", "Very", or "Fairly" satisfied with their family life. On the other hand, they could also express their varying levels of dissatisfaction, specifying to be "Fairly", "Very", or "Completely" dissatisfied with their family life.

VALUE LABELS

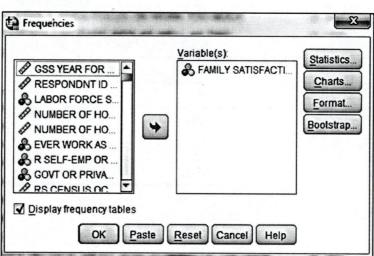

While the VALUE LABELS window reveals that respondents had numerous options to choose from, too many categories can also derail certain survey respondents. For instance, it is difficult to distinguish between being "Completely satisfied" in contrast to just "Very satisfied". In addition, too many options can also result in low frequencies for each category.

If we click on ANALYZE -> DESCRIPTIVE STATISTICS -> FREQUENCIES, the below illustrated window will open.

FREQUENCIES

By clicking OK, the following output will open. (Table 5)

Table 5 illustrates the frequencies for each category in addition to the given column percentages. A total of 1,263 respondents have indicated their level of family satisfaction. The majority of the respondents (43.4%) were "Very satisfied" with their family life, while 25% have indicated to be "Fairly satisfied". Other categories such as "Fairly dissatisfied", "Very dissatisfied", or "Completely dissatisfied" drew fewer respondents, resulting in much lower frequencies (i.e., 41, 22, and 10 respondents in total have opted for the aforementioned three categories). The researcher will have to collapse certain groups in order to create categories that have higher and more similar frequencies.

By selecting TRANSFORM -> RECODE INTO DIFFERENT VARIABLES, the students can collapse the above illustrated categories into fewer options.

TABLE 5

Family Satisfaction in General		Frequency	Percent	Valid Percent	Cumulative Percent
Valid	Completely satisfied	254	5.3	20.1	20.1
	Very satisfied	548	11.4	43.4	63.5
	Fairly satisfied	316	6.6	25.0	88.5
	Neither satisfied nor dissatisfied	72	1.5	5.7	94.2
	Fairly dissatisfied	41	.9	3.2	97.5
	Very dissatisfied	22	.5	1.7	99.2
	Completely dissatisfied	10	.2	.8	100.0
	Total	1263	26.2	100.0	
Missing	IAP	3518	73.0		
	Cannot choose	13	.3		
	NO ANSWER	26	.5		
	Total	3557	73.8		
Total		4820	100.0		

RECODE INTO DIFFERENT VARIABLES

	Name				Label	Value
541	singlpar				PARENTS CAN RAISE KIDS AS WELL AS TWO	{0, IAP}..
542	cohabok				OGETHER AS AN ACCEPTABLE OPTION	{0, IAP}..
543	divbest				E AS BEST SOLUTION TO MARITAL PROBLEMS	{0, IAP}..
544	fambudgt				OUPLES MONITOR BUDGET	{0, IAP}..
545	laundry1				HOUSEHOLD DOES LAUNDRY	{0, IAP}..
546	repairs1				HOUSEHOLD DOES SMALL REPAIRS	{0, IAP}..
547	caresik1				HOUSEHOLD CARES FOR SICK IN FAMILY	{0, IAP}..
548	shop1				HOUSEHOLD SHOPS FOR GROCERIES	{0, IAP}..
549	cooking1				HOUSEHOLD PREPARES THE MEALS	{0, IAP}..
550	rhhwork				ANY HOURS A WEEK DOES R SPEND ON HH WORK	{-1, IAP}..
551	sphhwork				ANY HOURS A WEEK DOES SPOUSE ON HH WRK	{-1, IAP}..
552	hhwkfair				G OF HH WORK BETWEEN R AND SPOUSE	{0, IAP}..
553	deckids				AKES DECISION ABOUT HOW TO BRING UP CHILDREN	{0, IAP}..
554	happy7				APPY R IS	{0, IAP}..
555	satjob7	Numeric	1	0	JOB SATISFACTION IN GENERAL	{0, IAP}..
556	satfam7	Numeric	1	0	FAMILY SATISFACTION IN GENERAL	{0, IAP}..
557	twoincs1	Numeric	1	0	BOTH MEN AND WOMEN SHOULD CONTRIBUTE TO INCOME	{0, IAP}..
558	earnshh	Numeric	1	0	HUBBY OR WIFE EARNS MORE DOLLARS	{0, IAP}..
559	SSFCHILD	Numeric	8	0	SAME SEX FEMALE COUPLE RAISE CHILD AS WELL AS MALE-FE...	{0, IAP}..
560	SSMCHILD	Numeric	8	0	SAME SEX MALE COUPLE RAISE CHILD AS WELL AS MALE-FEMA...	{0, IAP}..

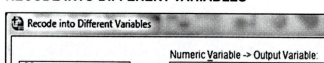

The RECODE INTO DIFFERENT VARIABLES dialogue box will open.

RECODE INTO DIFFERENT VARIABLES

TABLE 6

Original Variable "satfam7"	Newly recoded variable "Satfam7new"
"1" Completely Satisfied "2" Very Satisfied "3" Fairly Satisfied	"1" Satisfied
"4" Neither Satisfied nor Dissatisfied"	"2" Neutral/Neither Satisfied nor Dissatisfied
"5" Fairly Dissatisfied "6" Very Dissatisfied "8" Completely Dissatisfied	"3" Dissatisfied
"8" Cannot choose "9" No answer	"4" Do not know/No answer

Students will have the opportunity to select the variable they wish to recode by clicking on the arrow button to move the variable "satfam7" from the variable list located at the left to the NUMERIC VARIABLE -> OUTPUT VARIABLE dialogue box. Next, in the OUTPUT VARIABLE box, the researcher has to indicate the newly recoded variable's NAME and LABEL, as he/she decided on based on his/her preferences (i.e., "satfam7new" and its explanatory label "family satisfaction recoded").

The newly recoded variable will embrace only four categories. These new categories will also receive new codes: the "Satisfied" category will be coded as a "1" while the "Neutral" option will receive a code of "2". Moreover, respondents who were dissatisfied with their family life will be coded as a "3", and those who failed to answer the question will receive a code of "4".

After deciding on the new categories and their respective codes, the researcher will also have to collapse the original categories into the newly recoded classifications. For instance, the original categories "Completely", "Very", and "Fairly" illustrated in the VALUE LABELS table above will be collapsed into one single category. (See illustration in Table 6.)

Consequently, the recoding will take the following form:

> 1 thru 3 ->1
> 4->2
> 5 thru 7 ->3
> 8 thru 9->4

In order to enter the new values, we click on the OLD AND NEW VALUES button in the RECODE INTO DIFFERENT VARIABLES window illustrated below.

OLD AND NEW VALUES

We click on RANGE and in the first text box enter "1", followed by introducing "3" in the second text box. Next, we move to the NEW VALUE box and introduce "1". After clicking on ADD in the OLD -> NEW box, the following illustration will appear : "*1 thru 3->1*", indicating that the original variable's first three categories have been collapsed into solely one category.

In other words, the newly recoded variable encompasses the categories "Completely satisfied", "Very satisfied", and "Fairly satisfied" in one simple category. Similarly, we will collapse the original "Fairly dissatisfied", "Very dissatisfied", and "Completely dissatisfied" categories into one category recoded as a "3" (i.e., "*5 thru 7 ->3*"). Since the "Neither satisfied nor dissatisfied" category remains the same, we click on VALUE under the OLD VALUE box and enter the variable's original code ("4") followed by entering its new code in the NEW VALUE box ("2"). Next we click on ADD, and the OLD -> NEW box will illustrate the following expression: "*4->2*".

By clicking on CONTINUE, students will return to the original RECODE INTO DIFFERENT VARIABLE dialogue box. Next click on OK, and SPSS will carry out the recoding procedure. It is important for students to remember that besides the original variable "satfam7", a new variable will be created ("satfame7new"), shown at the end of the variable list. The researcher will have to click on the VALUE LABELS in the VARI-ABLE VIEW and manually enter the labels for each category. Thus, he/she will have to indicate that a code of "1" encompasses respondents who have pointed to be "Satisfied" with their family life as well as those who were "Neutral" ("2"), "Not satisfied" ("3"), or who failed to answer the question. (See below.)

VALUE LABELS

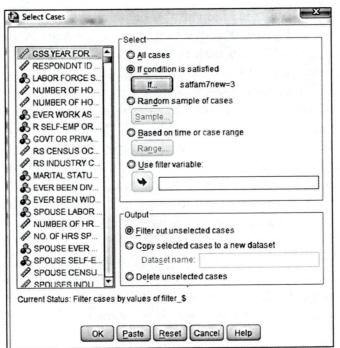

After having recoded the original variable, we can select the sample of respondents who have indicated to be dissatisfied with their family life.

Click on DATA -> SELECT CASES, and the SELECT CASES window (illustrated below) will open. The researcher can specify that solely respondents who have received a code of "3" shall be included in the subsequent analysis (i.e., "satfam7new=3").

SELECT CASES

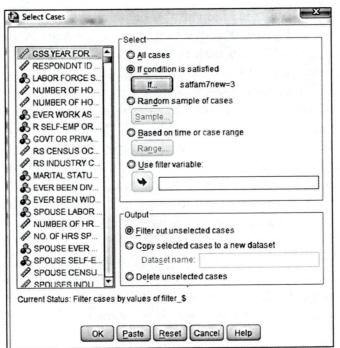

We click on OK, and SPSS will restrict the subsequent analysis to the subset of cases we have indicated above. Now, we click on ANALYZE -> COMPARE MEANS -> ONE-SAMPLE T TEST, and the below illustrated window will open.

ONE-SAMPLE T TEST

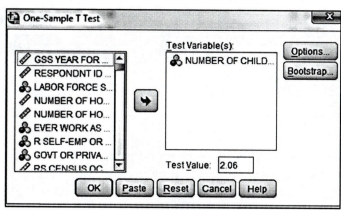

We select the interval-ratio variable "childs" (Respondents' Number of Children) in the TEST VARIABLE(S) box and enter the mean number of children an average American family had in 2012 in the TEST VALUE box[4]. SPSS will generate a ONE-SAMPLE STATISTICS (Table 7) and a ONE-SAMPLE TEST output (Table 8).

Table 7 displays some descriptive statistics, such as the number of respondents who have indicated to be dissatisfied with their family life (N=73), the average number of children they have (1.55), and the respective standard deviation.

If we contrast the sample average (1.55) presented in Table 7 with the average number of children American families had in 2012 (2.06), we observe that the latter is higher than the sample mean. Does this reveal that respondents who have indicated to be dissatisfied with their family life tend to have fewer children on average than the average American family? Is this difference between the sample mean (1.55) and the population mean (2.06) statistically significant, or is it the product of random chance?

TABLE 7

One-Sample Statistics				
	N	Mean	Std. Deviation	Std. Error Mean
Number of Children	73	1.55	1.764	.206

---

4 http://kff.org/global-indicator/total-fertility-rate/

TABLE 8

One-Sample Test						
	Test Value=2.06					
					95% Confidence Interval of the Difference	
	t	df	Sig. (2-tailed)	Mean Difference	Lower	Upper
Number of Children	−2.480	72	.015	−.512	−.92	−.10

In order to answer these questions and take an informed decision about whether we can reject the Null Hypothesis, we will have to look up the "Sig. (2-tailed)" value illustrated in the ONE-SAMPLE TEST output. The significance value of 0.015 displays the exact probability of the recorded difference between the sample mean of 1.55 children and the population mean of 2.06 to be caused by random chance alone. Since this probability is lower than our alpha level (0.05), we can reject the Null Hypothesis and conclude that Americans who are unsatisfied with their family life also tend to have fewer children than the average American family.

CHAPTER 6

Hypothesis Testing II: The Two-Sample Case

The four SPSS exercises presented in chapter 6 will gradually introduce students to concepts such as *equal variance* and *sample mean difference*. In addition, students will learn to carry out an Independent-Samples T Test and interpret the SPSS output of the *Levene's Test for Equality of Variances* and the *T-Test for Equality of Means*. While in chapter 5 (Hypothesis Testing I) we compared the difference between a sample mean and a population average, in this chapter we will contrast the mean differences between two population values.

In the first exercise, we will examine whether recent efforts to promote gender equality in employment opportunities have been successful at bridging the significant income differences between men and women. Does the gender gap's detrimental effect still persist in 2012, or has a cross-gender level playing field become a reality? We will use the INDEPENDENT-SAMPLES T TEST procedure in SPSS to probe the relationship between the variable "sex" (Respondent's Sex/Gender) and the variable "real-rinc" (Respondent's Income in Constant Dollars). Are the mean income differences in our sample "large enough" to reveal significant differences within the population of American men and women? Will we acknowledge the prevalence of a gender gap that continues to persist even in 2012? In Exercise 2 we will contrast African-Americans' income with the earning potential of Caucasian respondents. Despite many years of affirmative action, African-Americans still face a discernible income disadvantage vis-à-vis the population of Caucasian or even Latino backgrounds. Is this "racial wage gap" showing signs of dilution, or is it still intractable? In the third exercise, we will examine the relationship between religion and education, more in particular, whether religious fundamentalism can be eliminated through education. In Exercise 4 we will contrast the sample of women who have indicated to agree with abortion with women who condemn it. More specifically, we are interested to learn if more affluent/economically independent women would on average be "pro-choice" rather than "pro-life".

CONDUCTING A TWO-SAMPLE TEST IN SPSS

Exercise 1

The four exercises below use the 2012 General Social Survey (GSS) data set[1].

Serious efforts to promote gender equality in employment opportunities, status, working conditions, and pay have not yet bridged the significant income differences between men and women. The "gender gap" concept encapsulates this discrepancy[2].

In this exercise, we will use the INDEPENDENT-SAMPLES T TEST procedure in SPSS to probe the relationship between the variable "sex" (Respondent's Sex/Gender) and the variable "realrinc" (Respondent's Income in Constant Dollars). For this purpose, we are going to refer to a randomly selected sample of men and women contained in the 2012 General Social Survey (GSS) data set in the intent of finding out whether such a gender gap persists in 2012.

In order to carry out the Independent-Samples T Test, we click on ANALYZE -> COMPARE MEANS -> INDEPENDENT-SAMPLES T TEST.

INDEPENDENT-SAMPLES T TEST

The INDEPENDENT-SAMPLES T TEST dialogue box will open, allowing the researcher to select a TEST VARIABLE and a GROUPING VARIABLE from the list of variables located at the left. It is very important for students to remember that they can select *only* interval-ratio (or correctly coded ordinal) variables in the TEST VARIABLE(S) box. In this box, the researcher indicates the variable for which sample averages are calculated. Conversely, the GROUPING VARIABLE box allows for the selection of the nominal (or ordinal) variable that distinguishes between the two populations by selecting two representative samples. In our case, the interval ratio variable "realrinc" capturing

---

1   You can download the data from http://www3.norc.org/GSS+Website/Download/SPSS+Format/.
2   If you want to read more about "gender gap" and its implications, see the Global Gender Gap Report 2012, available from http://www3.weforum.org/docs/WEF_GenderGap_Report_2012.pdf.

respondents' income will be chosen in the TEST VARIABLE(S) box by clicking the top arrow between the variable list on the left and the TEST VARIABLE(S) window situated on the right.

While the selection of the test variable is quite straightforward, the grouping variable requires students to DEFINE GROUPS after selecting the given nominal or ordinal variable. When clicking on the arrow located between the variable list and the GROUPING VARIABLE box, automatically two question marks appear, and the DEFINE GROUPS option becomes active.

INDEPENDENT-SAMPLES T TEST

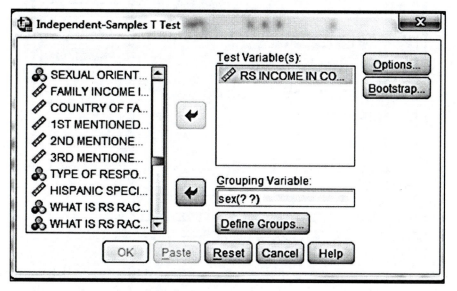

In order for students to correctly identify which cases go into which groups, it is essential to reopen the VALUE LABELS option in the data set's VARIABLE VIEW.

VALUE LABELS

The VALUE LABELS window illustrates the coding scheme applied for respondent's gender (variable "sex"). Consequently, when we click on the DEFINE GROUPS option, we know that group 1 will contain male respondents (coded as a "1", as we can see in the VALUE LABELS window), and females (coded as a "2") will be included in group 2. It is important for students to remember that they will have to manually enter the given codes for each group on every occasion they carry out an Independent-Samples T Test.

DEFINE GROUPS

After selecting the two groups so that we can proceed to make a distinction between our two samples, we click on CONTINUE, and the DEFINE GROUPS window will close. The INDEPENDENT-SAMPLES T TEST window will contain now the two values we have introduced in the GROUPING VARIABLE window.

TABLE 1

Group Statistics					
Respondents Sex		N	Mean	Std. Deviation	Std. Error Mean
RS Income in Constant $	Male	1357	36419.80	69146.650	1877.073
	Female	1470	19878.99	38188.495	996.033

By clicking OK, two output tables will be generated. The first table (GROUP STA-TISTICS) will present us with some basic descriptive statistics, such as the number of women and men in our sample, the average income each group earns, and their standard deviation. Thus, 1,470 women are represented in our sample, with a mean income of $19,878.99 and a standard deviation of 38,188.49. At the same time, the sample of men consists of 1,357 respondents, earning on average $36,419.80 with a standard deviation of 69,146.650.

There are two important observations at this point: (1) we see that women in our sample earn on average much less than men, and (2) the dispersion within the sample of men is much higher than the one we encounter among women (e.g., the standard deviation of men's income in our sample is higher than the one found among women).

The above mentioned observations raise two important questions:

(1) Is the difference in the sample means significant?

(2) Are the population variances equal?

Let us examine each question and its implications in detail.

(1) Is the difference in the sample means significant?

If the difference between the sample of men (N=1357, earning an average income of $36,419.80) and the sample of women (N=1470, earning an average income of $19,878.99) is statistically significant, we can infer from the observed differences in our sample to the population of women and men as whole. Consequently, we could say that such "large enough" differences in our samples reveal significant differences within the population of men and women and acknowledge the prevalence of a gender gap. However, the difference of $16,540.81 between the sample means of male and female incomes could have also been caused by random chance alone.

The Null Hypothesis reiterates exactly the latter assumption; that is, the difference between the sample averages was caused solely by random chance. In other words, the *population* of men and women are "no different" when it comes to their earning potential. Of course, the researcher is interested in rejecting the Null Hypothesis and demonstrating that such widespread discrepancies encountered within the sample averages (exactly $16,540.81) have not occurred by random chance, but, instead, they reveal the prevalence of a gender gap within the population represented by our samples.

(2) Are the population variances equal?

As we discussed in the first chapter, measures of dispersion, such as the standard deviation or variance, provide us with information on the data's dispersion. From a mathematic perspective, a distribution's variance is the standard deviation squared (i.e., raised at the power of two). Consequently, the two measures can be used interchangeably.

The GROUP STATISTICS table (Table 1) reveals that women's income in our sample has a lower dispersion than the income of men. We could reformulate the latter observation by stating that the variance (or standard deviation) among men's income is higher than the variance among women's earnings. In other words, given a normally distributed curve, among our 1,357 men in our sample, we will find that most of our respondents will "fall" further away from the calculated mean income of $36,419.80. Thus, we will find many below-average male respondents who earn considerably less than the above mentioned mean income (or eventually outliers earning zero [unemployed] or close to zero incomes [low income respondents or those on welfare]). At the same time, we will also have some men who surpass our sample average and will earn significantly higher incomes than $36,419.80.

Let us graphically illustrate the distribution of male respondents' income in our sample. First, we select the respondents in our sample who have indicated that they are men ("sex=1"). After clicking on DATA -> SELECT CASES -> IF CONDITION IS SATISFIED, we can select the variable "sex" and indicate that we would like to temporarily eliminate every respondent who has not received the code of "1" (i.e., females).

SELECT CASES

Following the selection of a sample of male respondents, we click on ANALYSE -> DESCRIPTIVES -> FREQUENCIES, and the FREQUENCIES window will open. In the VARIABLE(S) box, we select from the variable list on the left respondents' income ("realrinc") and click on the CHARTS button.

Under the FREQUENCIES: CHARTS window, we can select the type of chart we would like to display. Click on HISTOGRAM to display men's income as well as the normal curve.

FREQUENCIES: CHARTS

After clicking CONTINUE and OK, both FREQUENCIES: CHARTS and FREQUENCIES windows will close, and the following histogram will be illustrated.

GRAPH 1

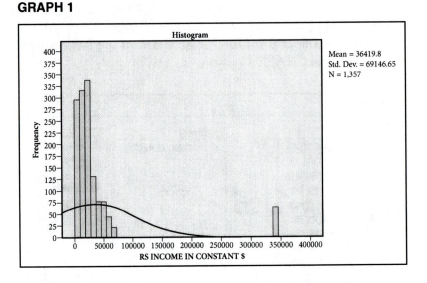

The high standard deviation illustrated in Table 1 reveals a high dispersion among men's scores. Our sample consists of numerous men indicating to be earning very low incomes, as the frequency in the "0" category displays close to 300 individuals at the poverty threshold. At the same time, we find an outlier cluster of close to 75 individuals indicating an income that reaches the $350,000 mark. All in all, the mean income of N=1357 male respondents is $36,419.80. However the dispersion among the scores— that is, the standard deviation—is very high.

It is interesting to contrast the histogram illustrating male respondents' income with the income distribution of female respondents. Similarly, we carry out an IF CONDITION (now indicating that the variable "sex" should be equal to "2") followed by the given histogram under the FREQUENCY command.

GRAPH 2

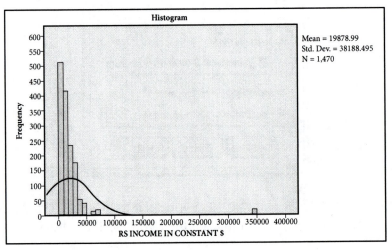

Graph 2 illustrates women's income distribution in our sample. While the mean income for our sample of 1,470 women is much lower than the average income for our male respondents, their standard deviation is also lower than men's. Graph 2 provides a clear depiction as to why women are poorer (i.e., lower average income) and "more similar" (i.e., lower standard deviation) than men.

If we contrast the two frequencies we observe that:

(1) The number of women earning no income or very low income is much higher than the number of men who are in the same situation. While we observed a frequency of close to 300 male respondents at or near the poverty threshold, this number is much higher for women (surpassing 500 respondents).

(2) The number of women who would earn above average or earn significantly more than the average is much lower as well. For instance, while the frequency for male outliers (high income earners) was 63, the same frequency for women in the $350,000 income bracket is 18. Moreover, the frequency of women earning close to $50,000 is much lower than that of men in the same income category.

Thus, we can conclude that:

(1) Female respondents in our sample earn on average less than male respondents.

(2) Women in our sample have a lower variance/standard deviation, revealing that our female sample is composed of "more similar" (or "similarly poorer") women while men are "more diverse". Our sample composition for men encompasses a relatively high number of (very) rich as well as (very) poor men.

Two questions then logically follow:

(1) Are these differences among the sample averages significant? Is the income difference in our sample average large enough to infer that a gender gap continues to persist in 2012?

(2) While the sample standard deviations revealed different dispersions (higher for men, lower for women) does such discrepancy in sample dispersion reflect differences in the variances within the population of men and women at large? In other words, is the population variance equal?

In order to answer these questions, we will have to interpret the results provided by the Independent-Samples Test output illustrated below (Table 2).

The Independent-Samples T Test is a "double test" composed of the Levene's Test for Equality of Variances and the T-Test for Equality of Means. In order to reject the Null Hypothesis, we will have to look up the Sig. (2-tailed) value in Table 2. If this value is lower than 0.05 (our alpha level), we reject the Null Hypothesis. On the contrary, if the value is higher than 0.05, we fail to reject the H_0. The Sig. (2-tailed) value is the exact probability of getting the observed difference of $16,540.81 (the difference between the men's and women's sample average income) if only random chance was operating.

TABLE 2

Independent Samples Test		Levene's Test for Equality of Variances		T-Test for Equality of Means					95% Confidence Interval of the Difference	
		F	Sig.	t	df	Sig. (2-tailed)	Mean Difference	Std. Error Difference	Lower	Upper
RS Income in Constant $	Equal variances assumed	87.538	.000	7.952	2825	.000	16540.809	2080.189	12461.967	20619.652
	Equal variances not assumed			7.784	2075.249	.000	16540.809	2124.967	12373.519	20708.099

If this probability is low, more exactly lower than our alpha of 5%, we reject the Null Hypothesis, asserting that the difference in sample means is not significant and random chance is the cause of it. In other words, the Sig. (2-tailed) probability value backs up our Null Hypothesis. The researcher's interest is to reject the latter and prove that the sample average differences are significant. The researcher can, thus, infer that such discrepancies between men's and women's income are characteristic of the whole population of American men and women.

However, when we look up our Sig. (2-tailed) value, we will find two rows containing two different significance and T-obtained values. The first row reveals a T-obtained score of 7.952 and a significance of 0.000, while the second row reflects a T-obtained value of 7.784 and a significance of 0.000. Thus, we first have to find out which significance and T-obtained values are correct—that is, if Equal Variances are assumed or not.

The Levene's Test of Equality of Variance will provide us with the given answer. SPSS assumes that the population variances are equal. Remember that women's scores (income) revealed a much lower dispersion than men's earnings. Hence, in our sample, the variances seemed to be unequal.

Nonetheless, we will have to carry out the Levene's Test in order to know if such assumption of equality of variance holds as well for the population of men and women. For this, we will have to look up the Sig. value in Table 1. If this value is lower than 0.05 (our alpha level), we reject the assumption of equal variances and move to the second row titled "Equal Variances Not Assumed". Thus, now we know that the correct T-obtained and Sig. (2-tailed) value we have to examine is 7.784 at a significance level of 0.000. The latter is much lower than 0.05, allowing us to reject the Null Hypothesis.

Conversely, if the sig. value next to the F ratio would have been higher than 0.05, we would fail to reject the Equal Variance assumption, and our correct T-obtained and significance values would be 7.952 and 0.000.

LEVENE'S TEST FOR EQUALITY OF VARIANCES

IF SIG.< 0.05 -> EQUAL VARIANCES ARE NOT ASSUMED -> LOOK UP CORRECT T-OBTAINED AND SIG. (2-TAILED) VALUES

IF SIG.>0.05 -> EQUAL VARIANCES ARE ASSUMED -> LOOK UP CORRECT T-OBTAINED AND SIG. (2-TAILED) VALUES

T-TEST FOR EQUALITY OF MEANS

IF THE SIG. (2-TAILED) VALUE < 0.05 -> REJECT THE NULL HYPOTHESIS

IF THE SIG. (2-TAILED) VALUE > 0.05 -> FAIL TO REJECT THE NULL HYPOTHESIS

In our case, the Equal Variances assumption was rejected since the first Sig. value next to the F ratio (0.000) was lower than 0.05. In addition, after looking up the correct Sig. (2-tailed) value, we are able to reject the Null Hypothesis of "no difference". Consequently, the gender gap's detrimental effects still persist in 2012, as we could generalize from the sample mean differences that such discrepancies between men's and women's incomes are prevalent in the population as well. The difference in the sample averages was not caused by random chance alone, and we can conclude that such discrepancies in income are statistically significant.

Exercise 2

The condition of African-Americans mirrors the gender gap problem. Despite many years of affirmative action, they still face a discernible disadvantage in the form of a racial wage gap vis-à-vis the population of Caucasian or even Latino backgrounds.

In this exercise we will compare the mean incomes of respondents who have indicated to be "black" with respondents who have identified themselves as "white". If the difference in the sample averages will be significant, we can reject the Null Hypothesis and conclude that the populations of African-Americans vis-à-vis the population of Caucasians greatly diverge in terms of their wage-earning positions in the American labor market. In other words, even in the year of 2012, equal opportunities despite one's race remain a utopia, and from a socioeconomic perspective African-Americans continue to be on the losing end.

After clicking on ANALYZE -> COMPARE MEANS -> INDEPENDENT-SAMPLES T TEST, the INDEPENDENT-SAMPLES T TEST window will open. The TEST VARIABLE(S) box will contain the interval-ratio variable for which SPSS will calculate the sample means.

INDEPENDENT-SAMPLES T TEST

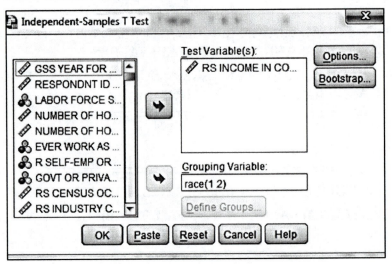

We are interested to compare "black" and "white" respondents' average income. The researcher has the opportunity to indicate which samples to study by selecting a nominal variable for the GROUPING VARIABLE box. Similarly to the first exercise, the researcher will have to manually enter the categories for each group. By clicking on the VALUE LABELS, the students can familiarize themselves with the nominal variable's coding in order to correctly indicate the two selected categories. The VALUE LABELS window illustrates that respondents' race encompasses three categories in addition to the inapplicable (IAP) option coded as "1". Thus, respondents who have indicated to be Caucasian have received a code of "1", African-Americans were coded as a "2", and all other races were included under category "3".

VALUE LABELS

TABLE 3

Group Statistics					
Race of Respondent		N	Mean	Std. Deviation	Std. Error Mean
RS Income in Constant $	White	2160	30156.26	60147.978	1294.178
	Black	411	17942.97	30192.136	1489.268

After taking a look at the given categories, we click on the DEFINE GROUPS option and indicate the code for each group that we would like to select. As already discussed, we will select "white" ("1") and "black" ("2") respondents representing our two individual samples.

Following the INDEPENDENT-SAMPLES T TEST command, two output tables are generated: (1) the GROUP STATISTICS and the (2) INDEPENDENT-SAMPLES T TEST tables.

In the first table, we will find some descriptive statistics, such as the total number of white respondents (N=2160), earning on average $30,156.26 with a standard deviation of 60,147.97. On the other hand, the number of African-Americans in our data set is much lower (N=411), earning also considerably less—that is, $17,942.97 on average. Is this difference of $12,213.29 in the sample means significant? Can we infer from this observed difference in our sample average the condition of the population at large? Is this difference large enough to allow us to conclude that the population of "white folks" in contrast to the population of African-Americans represented by our samples is wealthier?

We have noticed a similar situation in Exercise 1, where women in our sample earned on average less but had also a lower standard deviation (dispersion) than men. In Exercise 2 we see that our sample of 411 African-Americans earns on average less and also has a lower level of dispersion than Caucasian respondents. In order to see if the results in the GROUP STATISTICS table are significant, we will have to interpret the results illustrated in the INDEPENDENT-SAMPLES TEST table. (Table 4).

As discussed above, the Levene's Test for Equality of Variance precedes our T test for Equality of Means. Consequently, the first question we need to answer is related to the assumption of equality of variances. Is the assumption that the population's variances are equal accurate, or do we have to reject the former assumption?

If the Sig. value is less than 0.05, we are able to reject the Equal Variances assumption and state that the population variances are significantly different. This would confirm that the differences in standard deviation within our samples were significant. The 2,160 Caucasian respondents have on average a higher income but are also more dispersed around the mean. In other words, there is a higher likelihood for us to encounter among those 2,160 extremely rich Caucasians and also much poorer individuals. If we

TABLE 4

Independent Samples Test		Levene's Test for Equality of Variances		T-Test for Equality of Means							
									95% Confidence Interval of the Difference		
		F	Sig.	t	Df	Sig. (2-tailed)	Mean Difference	Std. Error Difference	Lower	Upper	
RS Income in Constant $	Equal variances assumed	27.435	.000	4.021	2569	.000	12213.291	3037.512	6257.071	18169.511	
	Equal variances not assumed			6.190	1139.631	.000	12213.291	1973.022	8342.127	16084.456	

contrast this result with the 411 Afro-Americans in our sample, we find that the former earns on average less but also has a lower variance. In other words, black people in our sample are poorer and more similar. Thus, we would find fewer outliers, and most of the respondents would be clustered around an average income of $17,942.97.

While SPSS assumes that the population variances are equal, we can reject the latter assumption because the first Sig. value (0.000) next to the F ratio is lower than 0.05. Thus, the correct T-obtained and significance values allowing us to reject the Null Hypothesis are 6.190 and 0.000 in the EQUAL VARIANCES NOT ASSUMED row.

The Null Hypothesis stated that the populations represented by our sample of Caucasian and African-Americans were "no different" concerning their earning potential. The difference we observed in the sample means is attributable solely to random chance. Our Research Hypothesis, on the other hand, presumed a significant difference and a racial wage gap in African-Americans' earning potential in contrast to the income of Caucasians. The Sig. (2- tailed) value indicates that the probability that solely random chance caused the wide discrepancy of $12,213.29 is close to zero (0.000), much lower than our required alpha level. Thus, we reject the Null Hypothesis and find such difference between the mean incomes in our two samples significant. Based on the 2012 GSS data, the racial wage gap that contributes to the unprivileged status of the African-American population fails to show signs of improvement.

Exercise 3

Over the past decade, the relationship between religion and education—more in particular, a question on whether religious fundamentalism can be eradicated through education—has sparked the interest of many researchers.[3]

---

3   See http://knowledge.sagepub.com/view/foundations/n308.xml.

In this exercise we will examine the relationship between Respondent's Highest Degree and the religious fundamentalism of his/her spouse. We would assume that respondents who have attained a higher degree will be less inclined to have partners with fundamentalist religious beliefs. While many studies suggest that fundamentalism cannot be countered with higher education, there is also a strong relationship between the level of schooling and the likeliness of a person to hold liberal values.

The VALUE LABELS window illustrates the religious beliefs of the respondent's spouse. More specifically, respondents were asked about their spouse's level of fundamentalism and were given three options: (1) Fundamentalist, (2) Moderate, (3) Liberal. In addition to these options, respondents could also indicate "not to know" the answer ("DK") or refuse to answer the question ("NA-EXCLUDED"), as well as inapplicable ("IAP").

VALUE LABELS

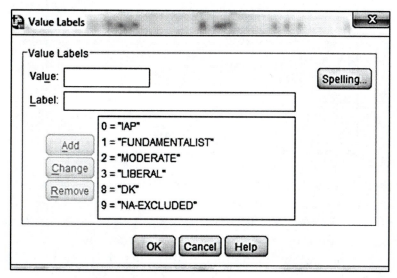

To carry out the Independent-Samples T Test, the researcher will select respondents who have indicated that their spouse shares "Fundamentalist" beliefs in contrast to respondents who have partners with "Liberal" views. The former category has received a code of "1" and the latter group a code numbered "3".

While the "spfund" variable distinguishes between the two populations, respondent's highest degree is the variable of interest for which sample averages will be calculated and compared. It is very important for students to remember that only interval-ratio variables can be selected in the TEST VARIABLE(S) box. *In some instances, however, we can replace the latter with correctly coded ordinal variables.*

The coding of Respondent's Highest Degree is illustrated in the VALUE LABELS window. Respondents who have indicated to have only some high school education received a code of "0". Similarly, high school graduates were coded as a "1", and junior college graduates received a code of "2", while respondents who have earned their bachelor's and graduate degrees were coded as a "3" and "4". The ascending nature of the coding, revealing gradually higher educational levels, allows us to use this ordinal variable as a TEST VARIABLE for which sample averages will be calculated. In other words, a higher average for a given sample reflects the former's higher educational standing, while a lower average shows us that the people in the group we are examining spent fewer years in the education system (i.e., have only primary and secondary education).

VALUE LABELS

It is important to note that if Respondent's Highest Level of Schooling would be coded in a descending inconsistent fashion—that is, a higher level of schooling would fail to reflect the given incremental increase in its code—students would have to recode the variable before conducting the Independent-Samples T Test.

After students have gotten acquainted with the nominal and ordinal variables' coding, we click on ANALYZE -> COMPARE MEANS -> INDEPENDENT-SAMPLES T TEST, and the INDEPENDENT-SAMPLES T TEST window will open.

INDEPENDENT-SAMPLES T TEST

TABLE 5

Group Statistics					
How Fundamentalist is Spouse Currently		N	Mean	Std. Deviation	Std. Error Mean
RS Highest Degree	Fundamentalist	566	1.56	1.172	.049
	Liberal	666	2.16	1.314	.051

By clicking on the first arrow between the variables list on the left and the TEST VARIABLE box on the right, we can select the variable "degree", for which sample means are calculated. Similarly, in the GROUPING VARIABLE box, we can define the two categories we selected in order to distinguish between the two populations. By clicking on the DEFINE GROUPS option, we select the respondents who have "Fundamentalist" ("1") or "Liberal" ("3") spouses. After we have indicated the nominal variable's categories, we click on OK, and two output tables will be generated.

The GROUP STATISTICS table reveals that 566 respondents have indicated to have a spouse sharing fundamentalist beliefs. In contrast, a total of 666 respondents have liberal partners. As expected, respondents who are married to a fundamentalist spouse have a mean score of 1.56, while the 666 respondents married to liberal spouses reveal a sample mean of 2.16 as their highest level of schooling.

Table 5 reveals that respondents who are married to a liberal spouse have also attained on average a higher level of education. A sample mean score of 2.16 reveals that respondents married to a partner with liberal views have on average a junior college (coded as a "2") or a bachelor's degree (coded as a "3"). In contrast, the people in the sample who have indicated to be married to a fundamentalist hold a high school (coded as "1") or junior college degree. Is this difference in the sample means significant?

Moreover, the standard deviation within the "Fundamentalist spouse" sample is lower (1.172) than the dispersion among respondents married to a liberal spouse (1.314). In other words, in the latter group we will encounter respondents whose degree ranges from high school to a bachelor's degree, while the former group's highest level of education will be clustered around the mean (i.e., encompassing a high school or junior college degree).

Are the population variances equal, and is the above displayed sample difference significant? In order to answer these two questions, we need to interpret the results illustrated in Table 6.

(1) Interpreting the Levene's Test for Equality of Variances

The first question we need to answer is related to the EQUAL VARIANCES assumption. Are equal variances assumed or not assumed? SPSS assumes that the population variances are equal. Based on the output in the GROUP STATISTICS table, the standard deviation within the "Liberal spouse" group was higher than the one encountered

TABLE 6

		Levene's Test for Equality of Variances		T-Test for Equality of Means						
									95% Confidence Interval of the Difference	
		F	Sig.	T	df	Sig. (2-tailed)	Mean Difference	Std. Error Difference	Lower	Upper
RS Highest Degree	Equal variances assumed	46.825	.000	−8.407	1230	.000	−.601	.072	−.741	−.461
	Equal variances not assumed			−8.485	1227.161	.000	−.601	.071	−.740	−.462

among respondents married to fundamentalist spouses. Consequently, from an educational perspective, the "Liberal spouse" sample was more diverse in terms of obtained highest degree than respondents married to a spouse who shared fundamentalist values. But does the difference in variances encountered within our samples undermine the assumption that the population variances are equal?

The first Sig. value next to the F ratio is 0.000, lower than our alpha level of 0.05, thus we can reject the Equal Variances assumption and state that the population variances are not equal. Rejecting the Equal Variances assumption will automatically provide us with the correct significance and T-obtained values.

(2) Interpreting the T-Test for Equality of Means

As discussed above, the correct T-obtained value is -8.485, and the significance is 0.000. The Sig. (2-tailed) value is lower than 0.05, allowing us to reject the Null Hypothesis.

The Null Hypothesis stated that there is "no difference" between the education levels of Americans married to liberal or fundamentalist spouses and the difference we observed in our sample means was caused by random chance. We were able to reject the H_0 and prove that the difference in education levels that we observed in our samples is significant. On average, Americans married to liberal spouses as opposed to those married to fundamentalist ones differ significantly in their level of education (i.e., they are better educated).

Exercise 4

In Exercise 4 we want to compare and contrast the sample of women who have indicated to agree with abortion with women who condemn it. More specifically, we are interested to learn whether more affluent/economically independent women would on average be "pro-choice" rather than "pro-life". Numerous researchers have sought to identify and describe the socioeconomic patterns that characterize women who opt for an abortion and/or advocate the primacy of a woman's choice. In this exercise we will compare the incomes of two samples (of pro-life and pro-choice women) to see if there is a significant difference between the two groups' earning potential.

In order to select solely female respondents, we click on DATA -> SELECT CASES -> IF CONDITION IS SATISFIED and indicate that the variable "sex" should equal the code of "2".

SELECT CASES

Following the IF CONDITION, we click on ANALYZE -> COMPARE MEANS -> INDEPENDENT-SAMPLES T TEST. The INDEPENDENT-SAMPLES T TEST window displays the two variables we have selected in the TEST VARIABLE(S) and GROUPING VARIABLE boxes.

INDEPENDENT-SAMPLES T TEST

TABLE 7

Group Statistics					
Abortion if Woman Wants for any Reason		N	Mean	Std. Deviation	Std. Error Mean
RS Income in Constant $	Yes	475	24433.32	46461.126	2131.783
	No	470	17471.36	38562.766	1778.768

TABLE 8

Independent Samples Test										
		Levene's Test for Equality of Variances		T-Test for Equality of Means					95% Confidence Interval of the Difference	
		F	Sig.	T	df	Sig. (2-tailed)	Mean Difference	Std. Error Difference	Lower	Upper
RS Income in Constant $	Equal variances assumed	4.020	.045	2.505	943	.012	6961.964	2779.131	1507.967	12415.960
	Equal variances not assumed			2.508	915.355	.012	6961.964	2776.421	1513.074	12410.853

We select "realrinc" (Respondents' Income in Constant Dollars) from the variable list, as the interval-ratio variable for which sample averages are calculated, and the nominal variable "abany", as the grouping variable distinguishing between our two samples. The latter variable has two categories, as respondents could indicate their support for abortion ("Yes", coded as a "1") or their disapproval ("No", coded as a "2"). Students will have to click on the DEFINE GROUPS option in order to indicate the scores of "1" and "2", determining which cases will go into which group category.

The GROUP STATISTICS table illustrates the two samples' descriptive statistics, while the INDEPENDENT-SAMPLES TEST output confirms the significance of the two-sample means' difference.

According to Table 5, we have a sample of 475 women who have indicated to be pro-choice with a mean income of $24,433.32 and a sample standard deviation of 46,461.126. On the other hand, 470 women in our sample are against abortion and display a pro-life attitude. The latter group earns on average $17,471.36 with a standard deviation of 38,562,766.

We immediately notice that women belonging to the "pro-choice" group outearn women in the "pro-life" category. Is this difference of approximately $6,962 between the two sample means significant, or was it caused by random chance alone?

Furthermore, the "pro-choice" sample reveals a higher dispersion, while the "pro-life" group is "more similar" in terms of their earning potential. Does the latter difference undermine the Equal Variances assumption in the population?

The Levene's Test for Equality of Variances displays a Sig. value of 0.045. Knowing that SPSS by default sets our alpha level at 5% (i.e., 0.05) and that 0.045 is lower than 0.05, we can reject the Equal Variances assumption and state that "equal variances are not assumed".

Nonetheless, the Sig. value comes extremely close to our alpha level. For instance, if we would want to decrease our alpha level to 1% (i.e., 0.01) or 0.1% (i.e., 0.001), in both cases the Equal Variances assumption could not be rejected. If students click on OPTIONS within the INDEPENDENT-SAMPLES T TEST window, they will have the opportunity to change their Confidence Interval from the SPSS default of 95% to 99% (resulting in an alpha of 1%) or 99.9% (resulting in an alpha of 0.1%).

INDEPENDENT-SAMPLES T TEST

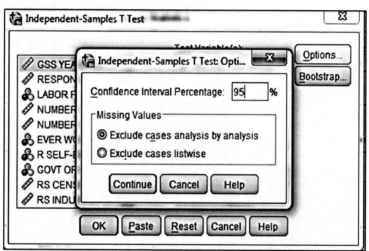

Certainly, the Confidence Level can also be decreased to 90%, resulting in an increased risk of 10%. In our case, due to the first Sig. value's proximity to our alpha level of 5%, the researcher can reject the Equal Variances assumption only at a 90% and 95% level of confidence. In contrast, if the researcher chooses to work with a very low risk factor and a higher confidence (i.e., 99% or 99.9%), the Equal Variances assumption has to be accepted. In each and every case, when one of the significance values comes very close to our alpha level, it is left at the discretion of the researcher to choose the appropriate confidence and alpha levels. Of course, a higher level of confidence (resulting in a lower level of risk) makes the rejection of the Null Hypothesis more difficult. Nonetheless, at the same time the risk of rejecting the Null Hypothesis incorrectly decreases significantly or virtually disappears with an alpha level of 1% or 0.1%. On the other hand, if the researcher's main purpose is to reject the Null Hypothesis, he/she might favor a lower confidence level (i.e., 90%) resulting in a much higher level of risk (i.e., 10%).

After we have rejected the Equal Variances assumption, the correct T-Test for Equality of Means values are 2.508 for our T-obtained and 0.012 for our Sig. (2-tailed) value. The latter is lower than 0.05, thus we can reject the Null Hypothesis. Therefore, in contrast to the "pro-life" advocates, the "pro-choice" American women demonstrate a higher earning potential. The sample mean differences are significant and cannot be attributed to random chance alone.

CHAPTER 7

Hypothesis Testing III: The Analysis of Variance (ANOVA)

In chapter 7 we will present the *Analysis of Variance* (ANOVA), one of the most prominent tests of statistical significance. ANOVA is very similar to the Two-Sample T Test presented in chapter 6. In contrast to HTII, ANOVA allows us to use independent variables with more than two categories. Consequently, in the five SPSS exercises discussed below, we will compare three or more population values. In the first two exercises, we will examine whether there is a significant difference in the number of children one decides to have based on one's level of education. We would assume that individuals with a graduate degree would be inclined to have fewer children than those without postgraduate education. The foregone assumption could apply, in particular, to women, who may be more inclined to postpone a life-impacting event, such as childbearing, for the sake of their career ambitions. We can entertain the question as to whether American women who have obtained undergraduate or graduate degrees have increasingly fewer children or opt out of motherhood altogether?

In the third exercise, we are testing the relationship between one's attitude toward the Bible and one's level of education. Do Americans who share different beliefs about the Bible also vary in their level of education? We would assume that individuals who have spent more years in higher education would also demonstrate an attraction toward rationalism and a concomitant demystification of the biblical precepts. In Exercise 4, we are interested to test the relationship between respondents' level of fundamentalism and their earning potential. Do Americans sharing different beliefs also vary in their earning potential?

In the last ANOVA exercise, we will probe the pervasiveness of racism in the American society. More in particular, we are interested to see if Caucasians with racist proclivities differ in terms of their earning potential vis-à-vis non-racist Americans. More specifically, we want to find whether Caucasians who have manifested a feeling of closeness ("Very close") to other races have, at the same time, a higher earning potential and an improved socioeconomic standing.

CONDUCTING A ONE-WAY ANOVA TEST IN SPSS

Exercise 1

The five exercises below use the 2012 General Social Survey (GSS) data set[1].

In this exercise we are interested to see whether there is a *significant difference* in the number of children one decides to have based on one's level of education. We would assume that individuals holding a graduate degree would be inclined to have fewer children than those who have not pursued postgraduate studies. The latter assumption could hold true especially in the case of women, where numerous statistics show that they are more inclined to postpone childbearing for the sake of pursuing their career ambitions.

Before we carry out the ONE-WAY ANOVA procedure, let us familiarize ourselves with the two variables we will use to complete this exercise. With the help of the VALUE LABELS option, we observe that respondents had five options to choose from when it came to indicating the highest degree they had earned. Respondents who have pointed to having obtained a graduate degree received a code of "4", while those who had not received their high school degree yet were coded as a "0".

In between these two categories, we will find respondents stating to have received a high school degree ("1"), a junior college degree ("2"), or a bachelor degree ("3"). To these four categories are, then, added three additional options: inapplicable ("7"), don't know ("8"), and not applicable ("9"). (See VALUE LABELS.)

VALUE LABELS

---

1   http://www.norc.org/Research/Projects/Pages/general-social-survey.aspx

In order to use the ONE-WAY ANOVA procedure, we click on **ANALYZE -> COMPARE MEANS -> ONE-WAY ANOVA as illustrated** below.

ONE-WAY ANOVA

File	Edit	View	Data	Transform	Analyze	Direct Marketing	Graphs	Utilities	Add-ons	Window	Help

	Name	Type				Label
29	age	Numeric	Reports ▸		
30	agekdbrn	Numeric	Descriptive Statistics ▸		
31	educ	Numeric	Tables ▸		
32	paeduc	Numeric	Compare Means ▸	Means...	
33	maeduc	Numeric	General Linear Model ▸	One-Sample T Test...	
34	speduc	Numeric	Generalized Linear Models ▸	Independent-Samples T Test...	
35	degree	Numeric	Mixed Models ▸	Paired-Samples T Test...	HEST DEGREE
36	padeg	Numeric	Correlate ▸	One-Way ANOVA...	RS HIGHEST DEGREE
37	madeg	Numeric	Regression ▸		RS HIGHEST DEGREE
38	spdeg	Numeric	Loglinear ▸		ES HIGHEST DEGREE
39	major1	Numeric	Neural Networks ▸		GE MAJOR 1
40	major2	Numeric	Classify ▸		GE MAJOR 2
41	DIPGED	Numeric	Dimension Reduction ▸		MA, GED, OR OTHER
42	SPDIPGED	Numeric	Scale ▸		E DIPLOMA, GED, OR OTHER
43	whenhs	Numeric	Nonparametric Tests ▸		RECEIVED HS DEGREE
44	whencol	Numeric	Forecasting ▸		RECEIVED COLLEGE DEGREE
45	sex	Numeric	Survival ▸		NDENTS SEX
46	race	Numeric	Multiple Response ▸		OF RESPONDENT
47	res16	Numeric	Missing Value Analysis...		OF PLACE LIVED IN WHEN 16 YRS OLD
48	reg16	Numeric	Multiple Imputation ▸		N OF RESIDENCE, AGE 16
49	mobile16	Numeric	Complex Samples ▸		RAPHIC MOBILITY SINCE AGE 16
50	family16	Numeric	Quality Control ▸		LIVING WITH PARENTS WHEN 16 YRS OLD
			ROC Curve...		

The ONE-WAY ANOVA window allows us to select one variable for the **DEPENDENT LIST** and one variable for the **FACTOR** box. While students often confuse which variable goes into which box, the rule of thumb is quite simple and straightforward. Like with Hypothesis Testing II (Independent-Samples T Test), the variable selected for the FACTOR box is the nominal (or ordinal) variable that distinguishes between the populations. In contrast, the DEPENDENT LIST box contains the interval-ratio (or correctly coded ordinal) variable for which sample means (averages) are calculated.

In this exercise, the interval-ratio variable "childs", capturing respondents' number of children, will be selected for the DEPEDENT LIST. Similarly, for the FACTOR box, we have chosen the variable "degree"—that is, respondents' highest level of obtained degree.

ONE-WAY ANOVA

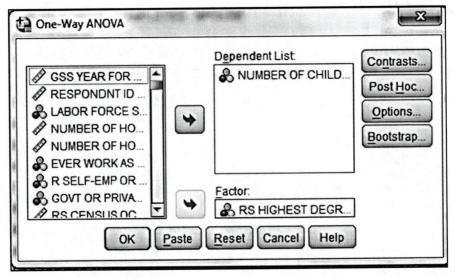

After selecting the two variables in the ensuing FACTOR and DEPENDENT LIST boxes, students will click on OPTIONS in order to select DESCPRIPTIVE in the STATISTICS rubric and, then, the MEANS PLOT rubric, graphically depicting the different sample means.

ONE-WAY ANOVA: OPTIONS

TABLE 1

Descriptives						
Number Of Children						
					95% Confidence Interval for Mean	
	N	**Mean**	**Std. Deviation**	**Std. Error**	**Lower Bound**	**Upper Bound**
LT High School	**607**	**2.92**	2.070	.084	2.75	3.08
High School	**2393**	**1.95**	1.613	.033	1.89	2.02
Junior College	**365**	**1.79**	1.404	.073	1.64	1.93
Bachelor	**925**	**1.47**	1.443	.047	1.37	1.56
Graduate	**523**	**1.51**	1.385	.061	1.39	1.63
Total	**4813**	**1.92**	1.666	.024	1.87	1.97

By clicking CONTINUE and OK, a new OUTPUT window will open, containing the DESCRIPTIVES and ANOVA tables in addition to the MEANS PLOTS graphical illustration.

The DESCPRIPTIVES table displays various summary statistics, such as the sample sizes, the sample averages, and the sample standard deviations. According to Table 1, a total of 4,813 respondents have indicated the highest degree they have earned. Among those, 607 respondents haven't yet or received solely their high school degree (N=2393). In addition, our data set includes 523 graduate degree and 925 bachelor degree holders, as well as 365 respondents who have received their junior college degree.

The second column reveals each sample's average number of children. Thus, the sample average for respondents who do not hold a high school degree is 2.92 (i.e., approximately three children on average). In contrast, respondents who hold a graduate degree have solely 1.51 children (i.e., one or two children), or in the case of those who have graduated with a bachelor degree, the sample mean further decreases to 1.47.

The MEANS PLOTS clearly illustrates this decreasing trend. While respondents' highest degree increases (X axis), at the same time, the average number of children is decreasing (Y axis). We find a slight increase of number of children only among those holding a graduate degree. Nonetheless, a sharp decreasing tendency is irrefutable. But, is this difference in the sample means *significant* or attributable solely to *random chance*?

GRAPH 1: MEANS PLOTS

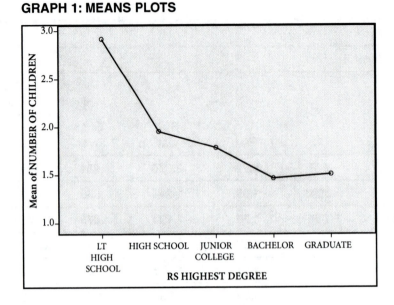

In order to carry out the Test of Statistical Significance, the researcher will need to interpret the results in the ANOVA table (Table 2). The latter includes "Between Groups" and "Within Groups" sum of squares, degrees of freedom, mean square estimates, F-ratio, and the significance value.

Similarly to Hypothesis Testing I and II, in order to reject the Null Hypothesis, we have to look up the Sig. value. If the latter is lower than our alpha level (i.e., 0.05), we can reject the H_0. Conversely, if the significance is higher than our risk factor, we fail to reject the Null Hypothesis.

IF THE SIG. VALUE < 0.05 -> REJECT THE NULL HYPOTHESIS

IF THE SIG. VALUE >0.05 -> FAIL TO REJECT THE NULL HYPOTHESIS

TABLE 2

ANOVA

Number Of Children

	Sum of Squares	df	Mean Square	F	Sig.
Between Groups	891.947	4	222.987	86.023	.000
Within Groups	12463.208	4808	2.592		
Total	13355.156	4812			

Because the Sig. value of 0.000 is lower than 0.05, we reject the Null Hypothesis and conclude that *the difference in the mean number of children among Americans holding degrees at various levels of the education system is statistically significant.*

Exercise 2

While the previous exercise included both female and male respondents, it would be interesting to see if the inverse relationship between the number of children and one's level of education holds true for the population of women. As already discussed above, childbearing has a stronger impact on women's careers, leading many women to opt out or have fewer children.

In order to select only female respondents, we click on DATA -> SELECT CASES -> IF CONDITION IS SATISFIED and identify the variable "sex" (Respondent's Sex) with the code of "2".

SELECT CASES

The VALUE LABELS window below illustrates the coding of the nominal variable "sex" capturing respondents' gender.

VALUE LABELS

Following the IF CONDITION, we click on ANALYZE -> COMPARE MEANS -> ONE-WAY ANOVA and reselect the same variables in the DEPENDENT LIST and FACTOR boxes.

ONE-WAY ANOVA

The DESCRIPTIVES table displays various summary statistics, including the sample sizes, the mean number of children, and their respective standard deviations. According to Table 3, the mean number of children is highest among the least educated women (i.e., an average number of 3.31 children). In contrast, women holding a bachelor degree tend to have only 1.52 children on average. This average is even lower for our sample of 287 female graduate respondents, who have the lowest mean when compared to the other four samples.

TABLE 3

Descriptives						
Number of Children						
					95% Confidence Interval for Mean	
	N	**Mean**	**Std. Deviation**	**Std. Error**	**Lower Bound**	**Upper Bound**
LT High School	**335**	**3.31**	2.065	.113	3.09	3.53
High School	**1346**	**2.08**	1.515	.041	2.00	2.16
Junior College	**213**	**1.94**	1.367	.094	1.75	2.12
Bachelor	**502**	**1.52**	1.431	.064	1.39	1.65
Graduate	**287**	**1.42**	1.218	.072	1.28	1.56
Total	**2683**	**2.05**	1.633	.032	1.99	2.11

This gradually decreasing trend is very well illustrated in the MEANS PLOT. In contrast to the first MEANS PLOT that also included male respondents, Graph 2 clearly illustrates that women with graduate, bachelor, or junior college degrees have on average fewer children than women who hold a high school degree or failed to graduate from high school. The sample mean gradually drops from 3.31 children to 1.42 on average.

GRAPH 2: MEANS PLOT

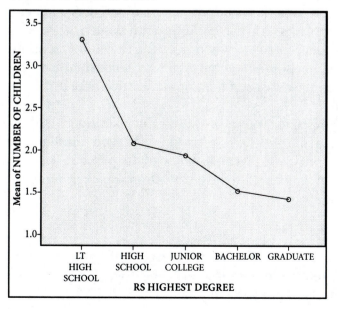

TABLE 4

ANOVA					
Number Of Children					
	Sum of Squares	df	Mean Square	F	Sig.
Between Groups	790.513	4	197.628	83.250	**.000**
Within Groups	6357.380	2678	2.374		
Total	7147.893	2682			

The only question left to answer is, is this difference in the sample means significant or attributable to random chance alone? Do American women who have obtained undergraduate or graduate degrees have increasingly fewer children or opt out of motherhood altogether? Are the trends identified within the five samples representative of the population at large?

The ANOVA table reflects a Sig. value of 0.000, less than 0.05, thus allowing us to reject the Null Hypothesis.

The Null Hypothesis states that the populations' means from which the samples are drawn are equal. In other words, irrespective of women's attained educational credentials, the average number of children they will opt to have will be equal. That is, women who have spent more years in higher education while receiving a postsecondary degree will tend to have an equal number of children as women who have completed only secondary schooling. In contrast to the H_0, *the Research Hypothesis says that at least one of the population averages needs to be different.* For instance, if American women with a graduate degree have fewer children than women with junior college degrees, the H_0 can be rejected. Moreover, when rejecting the Null Hypothesis, the ANOVA test will not distinguish the population mean(s) that is/are significantly different. Thus, we will only know that *at least one* of the population averages is different compared to the rest of the means.

Due to the significance value of 0.000, the likelihood that sole random chance caused the sample average difference is virtually zero, allowing us to reject the Null Hypothesis. Hence, we know that at least one of the populations' means differs from the rest of them and that women of diverse educational backgrounds tend to have different numbers of children.

Exercise 3

The Analysis of Variance is designed to be used with interval-ratio variables and nominal variables that include more than two categories. However, in some instances, we can replace the interval-ratio variable in our DEPENDENT LIST with a correctly coded ordinal variable. As we could observe in chapter 3 (Hypothesis Testing II), the variable "degree" (Respondent's Highest Degree) measures respondents' level of education, and due to its "correct" coding (i.e., increasing numerical codes capturing increased level of education), we can use the former to calculate sample averages. For instance, a lower sample average would reflect a lower level of education while a higher sample mean would reveal a higher standing.

We select the variable "bible", capturing respondents' attitude toward the Holy Script, as the nominal variable distinguishing between the different populations. Respondents could indicate whether they believed that the Bible was the "Word of God" (coded as a "1"), "Inspired Word" (coded as "2"), or a "Book of Fables" (coded as "3"), or they could also share a different perception ("Other" coded as a "4").

VALUE LABELS

By selecting the variable "degree" and the nominal variable "bible", we are testing the relationship between one's attitude toward the Bible and his/her level of education. We would assume that individuals who have spent more years in higher education would also develop a "different" (i.e., more rational) approach toward the Holy Script.

The Null Hypothesis states that the populations from which the samples are drawn are equal; that is, individuals sharing the idea that the Bible is the "Word of God" will have obtained a similar level of education as those who consider it to be solely a "Book of Fables" or an "Inspired Word". Our Research Hypothesis (H_1), in contrast to the Null Hypothesis, states that *at least one* of the population averages is different. In other words, at least one of our populations has a higher (or lower) level of education than the rest of the categories. For instance, rejecting the Null Hypothesis would imply that Americans who consider the Bible to be a book composed of fictional stories have a different (e.g., higher) level of education than those who consider it to be a Holy Script.

After we click on ANALYZE-> COMPARE MEANS-> ONE-WAY ANOVA, the ONE-WAY ANOVA window will open, allowing us to select the two variables mentioned above. The variable for which SPSS will calculate sample means will go into the DEPENDENT LIST box, while the variable that distinguishes between the populations will be selected for the FACTOR box.

ONE-WAY ANOVA

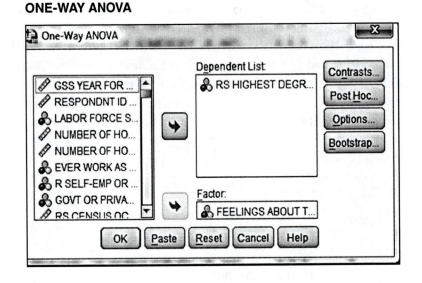

After selecting DESCRIPTIVES and a MEANS PLOT under OPTIONS, SPSS will generate two output tables and a graphical illustration of our sample means for each category we are comparing. The DESCRIPTIVES table (Table 5) illustrates the four different sample sizes and their respective level of education.

TABLE 5

Descriptives						
Rs Highest Degree						
					95% Confidence Interval for Mean	
	N	**Mean**	**Std. Deviation**	**Std. Error**	**Lower Bound**	**Upper Bound**
Word of God	1492	1.21	1.032	.027	1.16	1.26
Inspired Word	2185	1.82	1.225	.026	1.77	1.87
Book of Fables	1006	2.01	1.315	.041	1.93	2.09
Other	62	1.95	1.260	.160	1.63	2.27
Total	4745	1.67	1.230	.018	1.64	1.71

Thus, 1,492 respondents agreed that the Bible is the "Word of God" and revealed an average educational level of "1.21". In order to understand this numerical value, students should reopen the VALUE LABELS window for the coding of the variable "degree". A sample average of 1.21 indicates that respondents within this category have obtained on average either a high school (coded as "1") or junior college degree (coded as "2").

Interestingly, the highest level of cynicism (i.e., stating that the Bible is a "Book of Fables") arises from the most educated sample. The latter includes 1,006 respondents with a sample mean of 2.01, demonstrating that respondents who do not consider the Bible the "Word of God" hold, on an average, a college degree.

The MEANS PLOT reflects this gradual increase toward rationalism and a concomitant demystification that comes with better and more education.

MEANS PLOT

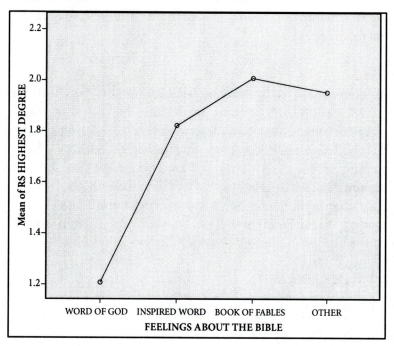

We find Bible devotees ("Word of God" and "Inspired Word") to have the two lowest sample averages, reflecting lower levels of education on average. In contrast, neutral supporters ("Other") and cynics ("Book of Fables") reveal a higher sample average, which means that postsecondary degrees are more frequent among them. The question then logically follows: is this difference *statistically significant*, or was it caused by random chance alone? In order to test for significance, students will need to interpret the Sig. value in the ANOVA table.

TABLE 6

ANOVA					
RS Highest Degree					
	Sum of Squares	df	Mean Square	F	Sig.
Between Groups	485.670	3	161.890	114.604	.000
Within Groups	6697.160	4741	1.413		
Total	7182.830	4744			

A significance level of 0.000 (< 0.05) allows us to reject the Null Hypothesis. At least one of the populations' means "is different" (H_1) and the sample mean difference illustrated in the DESCRIPTIVES table and the MEANS PLOT is not attributable to random chance alone. In other words, *Americans sharing different beliefs about the Bible vary in their level of education.*

Exercise 4

We could see in Exercise 3 that Americans' attitude toward the Bible varies with their obtained level of education. In this exercise, we are interested to test the relationship between respondents' level of fundamentalism and their earning potential.

The VALUE LABELS window reflects the variable's coding. The variable "fund" captures respondents' level of fundamentalism (i.e., How fundamentalist is RS currently?). Respondents could indicate a "Liberal" (coded as "3") or "Moderate" ("2") perception against displaying a higher level of fundamentalism ("1"). We find three additional categories: (1) Inapplicable, "IAP", coded "0"; (2) Do not know, "DK", coded "8"; and (3) Not applicable/ Excluded, "NA-Excluded", coded as "9".

VALUE LABELS

Understanding each variable's coding is a crucial step that precedes the Test of Statistical Significance. The latter will allow students to correctly identify the variables they should select for the DEPENDEPENT LIST and the FACTOR box in the ONE-WAY ANOVA window. The DEPENDENT LIST can contain solely interval-ratio or correctly coded ordinal variables while the FACTOR box should include a nominal (or ordinal) variable with more than two categories.

ONE-WAY ANOVA

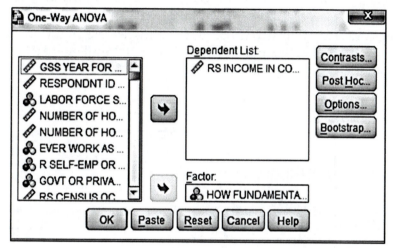

In this exercise, we have selected "realrinc", capturing respondents' income, as the variable for which sample averages will be calculated. On the other hand, the FACTOR box contains the variable "fund", distinguishing between respondents' moderate, liberal, and fundamentalist values. After selecting DESCRIPTIVES statistics and the MEANS PLOT under OPTIONS, we click on OK, and the following output tables and graphs appear.

TABLE 7

Descriptives						
RS Income in Constant $						
					95% Confidence Interval for Mean	
	N	**Mean**	**Std. Deviation**	**Std. Error**	**Lower Bound**	**Upper Bound**
Fundamentalist	653	20779.34	41774.708	1634.771	17569.29	23989.39
Moderate	1130	28898.15	57841.575	1720.683	25522.05	32274.25
Liberal	933	31949.87	63175.701	2068.280	27890.85	36008.90
Total	2716	27994.50	56556.632	1085.222	25866.55	30122.44

The first table (DESCRIPTIVES) displays various summary statistics, such as the sample sizes for each three categories, their respective sample averages, and standard deviations. A total of 2,716 respondents have indicated to have either liberal, moderate, or fundamentalist views with a total mean income of $27,994.50. Among those, 1,130 respondents pointed to share "Moderate" beliefs with an average earning potential of $28,898.15. A lower average income ($20,779.34) characterizes the sample of 653 individuals in our data set that claimed to be fundamentalists. In contrast, 933 liberal respondents greatly outearn both categories with a calculated mean income of $31,949.87.

The MEANS PLOT graphically illustrates the increasing trend on the Y axis (i.e., increased earning potential) for the three categories ("Fundamentalist", "Moderate", and "Liberal") depicted on the X axis.

MEANS PLOT

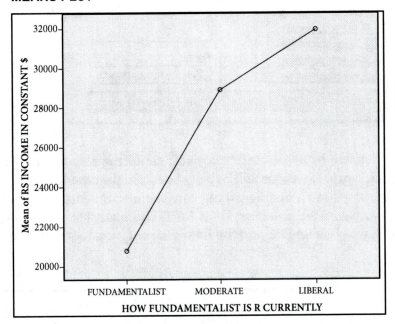

The "Liberal" sample reveals the highest earning potential, followed by respondents who have indicated "Moderate" beliefs, while the "Fundamentalist" group compares unfavorably to them. The difference between their sample average and the "Liberal" sample's mean income is approximately $11,170. When contrasted to the "Moderate" group, this discrepancy decreases to $8,118. Are these discrepancies in mean incomes significant, or were they caused by random chance alone? Can we infer from our observations and show that Americans sharing different beliefs also vary in their earning potential?

In order to answer these questions, students need to interpret the Sig. value in the ANOVA table (Table 8).

TABLE 8

ANOVA

RS Income in Constant $

	Sum of Squares	df	Mean Square	F	Sig.
Between Groups	49513702909.697	2	24756851454.849	7.778	.000
Within Groups	8634828266882.150	2713	3182760142.603		
Total	8684341969791.850	2715			

The Null Hypothesis states that the populations from which the samples are drawn are equal on the characteristic of interest. In other words, irrespective of their level of fundamentalism, the earning potential is similar among all Americans. In contrast, our Research Hypothesis says that *at least* one of the population means will be different; that is, fundamentalist or non-fundamentalist (moderate or liberal) Americans will have a different (i.e., lower or higher) income level.

The significance in the ANOVA table is lower than 0.05, thus we can reject the Null Hypothesis, proving that the mean difference in our samples was not caused by random chance but is significant. *Consequently, at least one of the population means is different, and Americans' earning potential varies based on their level of fundamentalism.*

Exercise 5

In the last ANOVA exercise, we will test for the pervasiveness of racism in the American society. More in particular, we are interested to see if Caucasians who display racist tendencies differ in terms of their earning potential vis-à-vis non-racist Americans.

The variable "closeblk" captures the levels of emotional closeness respondents feel vis-à-vis African-Americans. When we click on ANALYZE -> DESCRIPTIVE STATISTICS -> FREQUENCIES, the table illustrated below (Table 9) presents us with the given ordinal variable's output. The variable "closeblk" has altogether nine categories that measure the level of closeness on an ordinal scale from 1 to 9. The scales of "1" to "3" reflect a level of emotional detachment ("Not at all close"), while the codes of "4" to "6" reveal a neutral level of attachment. In contrast, respondents could also indicate to feel "Very Close" to African-Americans, encompassed in the 7–9 numerical intervals.

The ANOVA test is designed for nominal (or ordinal) variables with more than two categories, but in this case, when variables encompass up to nine categories, the variable becomes inadequate to carry out the Analysis of Variance. Consequently, before we conduct a One-Way ANOVA test, some categories will have to be collapsed into groups that have similar sample sizes. In order to accomplish the latter, the researcher can use the RECODE INTO DIFFERENT VARIABLES command, found under the TRANSFORM menu bar.

TABLE 9

How Close Feel to Blacks		Frequency	Percent	Valid Percent	Cumulative Percent
Valid	NOT AT ALL CLOSE	125	3.4	5.2	5.2
	2	55	1.5	2.3	7.5
	3	103	2.8	4.3	11.8
	4	118	3.2	4.9	16.7
	NEITHER ONE OR THE OTHER	1076	29.1	44.9	61.6
	6	216	5.8	9.0	70.6
	7	328	8.9	13.7	84.3
	8	146	3.9	6.1	90.4
	VERY CLOSE	231	6.2	9.6	100.0
	Total	2398	64.8	100.0	
Missing	IAP	1280	34.6		
	DK	16	.4		
	NA	6	.2		
	Total	1302	35.2		
Total		3700	100.0		

RECODE INTO DIFFERENT VARIABLES

Once we have clicked on the above illustrated command, the RECODE INTO DIFFERENT VARIABLES dialog box will open.

RECODE INTO DIFFERENT VARIABLES

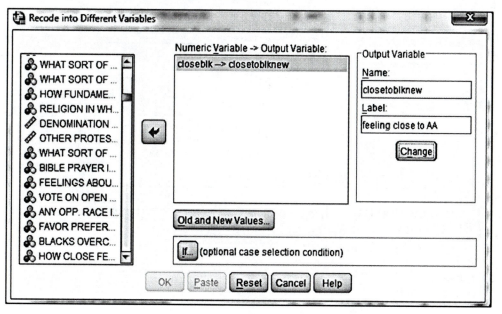

We select the variable we wish to recode from the variable list located at the left by clicking on the arrow sign to move the variable "closeblk" to the NUMERIC VARI-ABLE -> OUTPUT VARIABLE box. In the OUTPUT VARIABLE window, the students have the opportunity to introduce a name for the output variable they are recoding, in addition to an explanatory label. We enter the variable name "closetoblknew" and an explanatory label (i.e., "feeling close to AA"). After specifying the new output variable, we click on CHANGE and the output variable will appear in the NUMERIC VARIABLE -> OUTPUT VARIABLE box. At this point, the researcher will have to click on the OLD AND NEW VALUES button to open the RECODE INTO DIFFERENT VARIABLES: OLD AND NEW VALUES dialogue box. Select the RANGE option on the left-hand side and reexamine the DESCRIPTIVES output in Table 7.

Accordingly, we could recode the given variable into three groups capturing respondents' feelings toward African-Americans. The original nine categories will be collapsed into three new categories: (1) Not close at all, (2) Neutral, and (3) Feeling very close.

Thus we will click on RANGE and introduce the first and last value for the first cat-egory ("Not close at all"):

- Range: enter "1" through "3", followed by clicking on the ADD button. The expression "1 thru 3->1" will appear in the OLD -> NEW text box. (See below.)

RECODE INTO DIFFERENT VARIABLES: OLD AND NEW VALUES

Similarly to the first interval, we enter the subsequent categories. The OLD -> NEW text box will contain the following recoding:

1 thru 3-> 1
4 thru 6-> 2
7 thru 9-> 3

At this point two observations are essential. First, it is important for students to select the RECODE INTO DIFFERENT VARIABLES as opposed to the RECODE INTO SAME VARIABLES command under the TRANSFORM menu bar option. While the former creates a new variable based on the coding of the original variable, the latter command deletes the original variables (similar to "Save" versus "Save as" in Microsoft Word).

Second, after introducing the new intervals (ranges), by clicking CONTINUE and OK, the new recoded variable will be created. Students often forget to click on VARIABLE VIEW and under the VALUE LABELS options enter the Label for each recoded variable's category. (See VALUE LABELS window below.)

VALUE LABELS

Following a successful recoding of the above illustrated ordinal variable, we have created a variable that encompasses three instead of the original ten categories. By clicking on ANALYZE -> COMPARE MEANS -> ONE-WAY ANOVA, the ONE-WAY ANOVA window will appear. The DEPENDENT LIST contains the interval-ratio level variable "realrinc" for which sample averages will be calculated. On the other hand, the FACTOR box includes the new recoded variable capturing respondents' feelings toward African-Americans.

ONE-WAY ANOVA

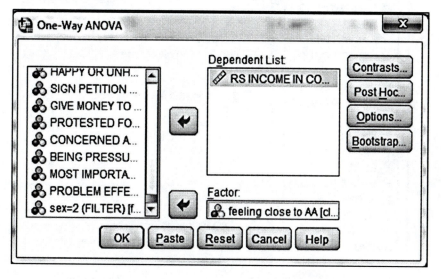

Because the original purpose of the test was to shed light on racism among Caucasian respondents of disparate socioeconomic backgrounds, we need to carry out an IF CONDITION before completing the ANOVA test.

VALUE LABELS

We click on DATA -> SELECT CASES-> IF CONDITION IS SATISFIED and indicate that only Caucasian respondents constitute the selected sample (i.e., race=1, see VALUE LABELS for the variable "race").

SELECT CASES

Following the IF CONDITION, we can return to the ONE-WAY ANOVA test and click on OK. Two output tables will be generated (DESCRIPTIVES and ANOVA) in addition to a MEANS PLOT.

The DESCRIPTIVES table displays the three sample sizes and respective sample means and standard deviations. Among Caucasian respondents, 145 have indicated to feel "Not close at all" to African-Americans, with a mean income of $29,940.53. In contrast, 424 have pointed to feeling "Very close", while also revealing the highest earning capacity compared to the "Not close at all" and "Neutral" groups. Is this difference in the sample means significant, or is it attributable solely to random chance? Do Caucasians who feel "Very close" to other races also reveal a higher earning potential and an improved socioeconomic standing?

TABLE 10

Descriptives						
RS Income in Constant $						
					95% Confidence Interval for Mean	
	N	**Mean**	**Std. Deviation**	**Std. Error**	**Lower Bound**	**Upper Bound**
Not close at all	145	29940.53	61128.919	5076.480	19906.49	39974.57
Neutral	837	31055.07	59789.366	2066.623	26998.69	35111.45
Very close	424	33400.34	68493.215	3326.325	26862.16	39938.53
Total	1406	31647.38	62640.845	1670.571	28370.30	34924.46

TABLE 11

ANOVA					
RS Income in Constant $					
	Sum of Squares	**df**	**Mean Square**	**F**	**Sig.**
Between Groups	2018975790.643	2	1009487895.321	.257	.773
Within Groups	5511026085571.130	1403	3928029996.843		
Total	5513045061361.770	1405			

The ANOVA table points to a Sig. value of 0.773, higher than 0.05, thus we fail to reject the Null Hypothesis. The latter stated that the populations from which the samples were drawn are equal on the characteristic of interest. In other words, white Americans who feel "Very close" to African-Americans have a similar earning potential as Caucasians who feel to be distant or do not show any attitude about the former race. Consequently, the researcher's attempt to demonstrate a relationship between racism and socioeconomic status has been unsuccessful.

CHAPTER 8

Hypothesis Testing IV: The Chi-Square Test and Measures of Association

Chapter 8 introduces one of the most popular SPSS tests, the Chi-Square Test. Its popularity owes largely to its simplicity: the Chi-Square Test waives the requirement of interval-ratio or correctly coded ordinal level variables, and it operates with variables measured at the nominal level. Therefore, Hypothesis Testing IV, in contradistinction to HT I, II, and III, does not juxtapose population values but offers insights into the relationship between two nominal variables. More in particular, it probes whether two variables are *dependent* or *independent*. On the other hand, *Measures of Association* serve as a *complementary* procedure to tests of statistical significance. They measure the strength and direction of the relationship between two variables. Exercises 1–3 will introduce three measures of association (*Phi*, *Cramer's V*, and *Lambda*) in addition to carrying out the Chi-Square Test to prove *dependence* or *independence* between variables.

In the first exercise, we would like to describe the relationship between the nominal variable "colrac" (that is, respondents' attitudes toward racist teachers) and the respondents' race. More specifically, we want to prove that one's opinion on permitting racists to teach is *dependent* on the variable "race". In the second exercise, we will evaluate the extent to which higher levels of education influence one's opinion on lesbian and gay rights. We assume that the better educated respondents would also exhibit stronger support and tolerance toward non-heterosexual individuals. In Exercise 3, we want to see if women's and men's attitudes on gun control issues differ. In other words, is the variable "gunlaw" (Favor or Oppose Gun Permits) *dependent* on the variable "sex" (Respondent's Gender), or are the two variables *independent*?

CONDUCTING A CHI-SQUARE TEST IN SPSS

Exercise 1

The exercises and illustrations below use the 2012 GSS data set[1].

In this exercise we would like to shed light on the relationship between the nominal variable "colrac", capturing respondents' attitudes toward racist teachers and respondents' race.

More specifically, we ask two research questions:

- Are the two variables dependent or independent?
- Which race is most likely to agree with "allowing racists to teach"?

Table 1 illustrates the frequency and percentage (Valid and Cumulative) of respondents who would agree/disagree with allowing a "racist to teach". Accordingly, a total of 2,356 respondents have answered the question "Would you allow a racist to teach?", and oddly 54.8% (more exactly, 1,292) agreed that racism should not be stigmatized and forbidden in the school system. Conversely, 45.2% of the respondents would "not allow" a racist to teach.

In order to carry out a Chi-Square Test, we click on ANALYZE -> DESCRIPTIVE STATISTICS -> CROSSTABS.

CHI-SQUARE TEST

TABLE 1

Allow Racists to Teach		Frequency	Percent	Valid Percent	Cumulative Percent
Valid	ALLOWED	1292	34.9	54.8	54.8
	NOT ALLOWED	1064	28.8	45.2	100.0
	Total	2356	63.7	100.0	
Missing	IAP	1280	34.6		
	DK	51	1.4		
	NA	13	.4		
	Total	1344	36.3		
Total		3700	100.0		

The CROSSTABS dialogue box will appear, allowing us to select the two nominal variables.

CROSSTABS

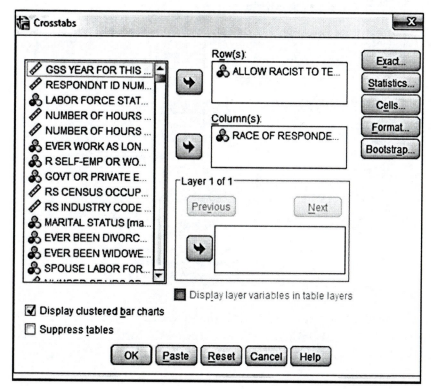

We move the independent variable from the variable list into the COLUMN(S) box and the dependent variable into the ROW(S) box. It is important for students to remember that *rows and columns cannot be used interchangeably*. Thus, the researcher always takes a conscious decision about the "cause" (independent variable) and the "effect" (dependent variable) when studying two variables' dependence.

After selecting the above mentioned variables, we click on STATISTICS, and the CROSSTABS: STATISTICS window will appear. The latter allows us to select the Chi-Square Test and Measures of Association, such as the Phi, Cramer's V, and Lambda. Measures of Association serve as a *complementary test* to tests of statistical significance. They measure the strength and direction of the relationship between two variables.

CROSSTABS: STATISTICS

In addition, we are also interested to display COLUMN PERCENTAGES in the SPSS generated bivariate table. By clicking on CELLS in the original CROSSTABS window, the CROSSTABS: CELL DISPLAY window will open.

CROSSTABS: CELL DISPLAY

Crosstabs: Cell Display

Counts
- ☑ Observed
- ☐ Expected
- ☐ Hide small counts
  - Less than 5

z-test
- ☐ Compare column proportions
  - ☐ Adjust p-values (Bonferroni method)

Percentages
- ☐ Row
- ☑ Column
- ☐ Total

Residuals
- ☐ Unstandardized
- ☐ Standardized
- ☐ Adjusted standardized

Noninteger Weights
- ◉ Round cell counts
- ◯ Round case weights
- ◯ Truncate cell counts
- ◯ Truncate case weights
- ◯ No adjustments

[Continue] [Cancel] [Help]

Click on CONTINUE and OK, and the four output tables (illustrated below) in addition to a bar chart will appear. Table 2 displays the bivariate Crosstabulation of the independent ("race": Race of Respondent) and dependent variables ("colrac": Allow Racists to Teach).

TABLE 2

Allow Racist to Teach * Race of Respondent Crosstabulation						
			Race of Respondent			
			White	Black	Other	Total
Allow Racist To Teach	ALLOWED	Count	1292	191	105	1588
		% within Race of Respondent	54.8%	42.4%	40.9%	51.8%
	NOT ALLOWED	Count	1064	260	152	1476
		% within Race of Respondent	45.2%	57.6%	59.1%	48.2%
Total % within Race of Respondent		Count	2356	451	257	3064
			100.0%	100.0%	100.0%	100.0%

Each cell displays the *observed frequency* and the column percentages for the given observation. For instance, 54.8% of Caucasian respondents (or 1,292) have agreed to "allow racists to teach". In contrast, the bulk of African-American and "Other" races disagree with the overwhelming majority of "White" respondents. Thus, 57.6% (or 260) "Black" respondents have indicated that racists should not be allowed to teach. Similarly, 59.1% of "Other" races agreed that racism should not be tolerated among teachers. As a general conclusion, Caucasians were more inclined to "Allow" racists to teach in contrast to blacks or other races. The bar chart below displays the above mentioned tendency.

GRAPH 1

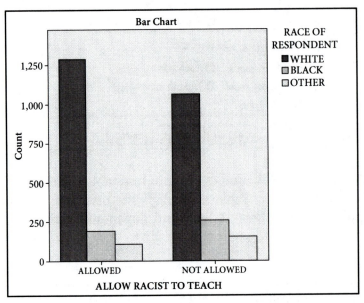

Are the two variables dependent or independent? *The Null Hypothesis states that the variables are independent* and the relationship we could observe in the Crosstabulation table has been caused by random chance alone. In contrast, the Research Hypothesis states that the variables are dependent. According to Healey and Prus (2010), "two variables are independent if the classification of a case into a particular category of one variable has no effect on the probability that the case will fall into any particular category of the second variable"[2]. In other words, whether an individual is Caucasian, African-American, or "Other" has no bearing on one's opinion about "allowing racists to teach". However, Table 2 illustrates Caucasians' higher likelihood to tolerate such racism among teachers in contrast to blacks and other races that were clearly against it.

Ultimately as with any other Test of Significance, the decision to reject the Null Hypothesis is illustrated in Table 3 (CHI-SQUARE TESTS).

---

2   Joseph F. Healey and Steven G. Prus (2010), *Statistics, A Tool for Social Research.*

TABLE 3

Chi-Square Tests			
	Value	Df	Asymp. Sig. (2-sided)
Pearson Chi-Square	37.172	2	.000
Likelihood Ratio	37.246	2	.000
Linear-by-Linear Association	33.489	1	.000
N of Valid Cases	3064		

The Sig. (2-sided) value is 0.000, less than 0.05, thus allowing us to reject the Null Hypothesis. Consequently, the variable "race" and one's opinion on allowing racists to teach are *dependent*. The significance value, as in all other Tests of Statistical Significance, captured the exact probability of getting the observed frequencies illustrated in Table 1, if only random chance was operating.

Thus we can conclude that there is a *statistically significant relationship* between the nominal variable "race" and the nominal variable "colrac", capturing respondents' attitudes towards racist teachers.

While the Chi-Square Test allows us to test if two variables are dependent, in order to answer the second set of questions related to the strength and the direction of the relationship, students will need to use Measures of Association. The latter provides information complementary to the Chi-Square Test. It is important for students to remember that Measures of Association shall be used solely if the Null Hypothesis of independence has been rejected. *In other words, we are interested in the strength of two variables' association if their relationship proved to be significant.*

The Measures of Association (MOA) are displayed in Table 4 (SYMMETRIC MEASURES) and Table 5 (DIRECTIONAL MEASURES). The researcher selects the preferred Measure of Association in the CROSSTABS: STATISTICS window under the NOMINAL label.

There are three MOAs we will discuss in this chapter:

(1) Phi
(2) Cramer's V
(3) Lambda

The first two measures (Phi and Cramer's V) reported in Table 4 are symmetric measures; that is, irrespective of which variable we choose as the dependent or independent, the same value will be displayed. While we can make use of Phi for nominal variables that have only two categories (i.e., 2 × 2 matrix with two rows and two columns), Cramer's V is a measure of association for variables with more than two categories.

TABLE 4

Symmetric Measures		Value	Approx. Sig.
Nominal by Nominal	Phi	.110	.000
	Cramer's V	.110	.000
N of Valid Cases		3064	

The value of Phi and Cramer's V is between zero and one. The closer the value comes to zero, the weaker the relationship.

More specifically, if the value is:

- Between 0.00 and 0.10, we speak of a weak relationship between the two variables
- Between 0.11 and 0.30, the relationship is moderate
- Between 0.30 and 1.00, the relationship is strong

In our case, the independent variable "race" has three columns ("White", "Black", and "Other") while the dependent has only two ("Allowed", "Not Allowed"). Thus, we have a 3 × 2 matrix (three rows and two columns), and the correct Measure of Association we need to look up in Table 4 is Cramer's V.

The latter reveals a weak association between the variable "race" and one's opinion on allowing racists to teach.

Table 5 (DIRECTIONAL MEASURES) displays three values for Lambda. It is important for students to remember that Lambda is an *asymmetric measure*, which means that its value will change according to the value of the dependent variable. In other words, if we would swap the independent with the dependent variable, the value for Lambda would change. Moreover, Lambda belongs to the group of measures based on the *proportional reduction in error (PRE)* logic. The latter measures the percentage of predicting errors the researcher can circumvent while forecasting the value of the dependent variable with the help of the independent.

The correct value for Lambda is 0.079 (our dependent variable was "colrac", "allowing racists to teach"), illustrated in the second row. A Lambda of 0.079 indicates that the researcher would make 7.9% fewer errors in predicting respondents' opinions on allowing racists to teach (the dependent variable) on the basis of their race ("race" as our independent variable) rather than without the latter.

Exercise 2

In this exercise, we are interested to see if Americans are showing greater tolerance and acceptance toward the members of the lesbian and gay community. While public opinion is still divided, younger generations today exhibit a much stronger approval of gays and lesbians. Thus, we would like to test how higher levels of education influence

TABLE 5

Directional Measures			Value	Asymp. Std. Error[a]	Approx. T[b]	Approx. Sig.
Nominal by Nominal	Lambda	Symmetric	.053	.012	4.373	.000
		ALLOW RACISTS TO TEACH Dependent	.079	.017	4.373	.000
		RACE OF RESPONDENT Dependent	0.000	0.000	[c]	[c]

TABLE 6

Allow Homosexuals to Teach		Frequency	Percent	Valid Percent	Cumulative Percent
Valid	ALLOWED	2645	54.9	85.8	85.8
	NOT ALLOWED	437	9.1	14.2	100.0
	Total	3082	63.9	100.0	
Missing	IAP	1669	34.6		
	DK	50	1.0		
	NA	19	.4		
	Total	1738	36.1		
Total		4820	100.0		

one's opinion on homosexuals' rights. The variable "colhomo" captures respondents' feelings and tolerance toward "allowing homosexuals to teach". Table 6 illustrates the number and percentage of respondents agreeing or disagreeing with homosexual teachers. The overwhelming majority of respondents expressed their agreement (85.8%), while 14.2% of those who have indicated a preference opted to ban homosexuals from the possibility of a teaching career.

Are more educated respondents more or less inclined to "allow" or "not allow" homosexuals to teach? Moreover, is the variable that captures respondents' highest level of education dependent or independent from one's opinion and tolerance toward gays and lesbians? If the relationship is significant, how strong is the association between the two variables?

In order to answer these questions, we will have to carry out a Chi-Square Test. By clicking on ANALYZE -> DESCRIPTIVE STATISTICS -> CROSSTABS, the CROSSTABS window will open.

CROSSTABS

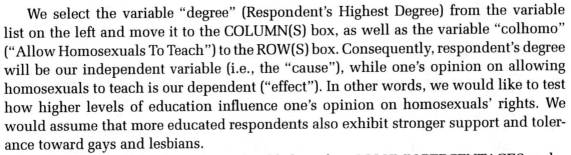

We select the variable "degree" (Respondent's Highest Degree) from the variable list on the left and move it to the COLUMN(S) box, as well as the variable "colhomo" ("Allow Homosexuals To Teach") to the ROW(S) box. Consequently, respondent's degree will be our independent variable (i.e., the "cause"), while one's opinion on allowing homosexuals to teach is our dependent ("effect"). In other words, we would like to test how higher levels of education influence one's opinion on homosexuals' rights. We would assume that more educated respondents also exhibit stronger support and tolerance toward gays and lesbians.

Similarly to Exercise 1, students should also select COLUMN PERCENTAGES under CELL DISPLAY and MEASURES OF ASSOCIATION under STATISTICS. After clicking CONTINUE and OK, the output below will be produced.

The Crosstabulation table (Table 7) displays two rows and five columns (2 × 5 matrix). While only a total of 14.2% of respondents expressed their intolerance toward homosexual teachers, an overwhelming majority of the respondents from this category have also obtained low levels of education. For instance, 31.6% of respondents who oppose gay teachers have only little high school education, followed closely by 17.6% who have solely a high school degree. In contrast, an overwhelming majority of highly educated respondents, who have obtained at least a bachelor or graduate degree, expressed

TABLE 7

Allow Homosexual to Teach * Rs Highest Degree Crosstabulation								
			RS Highest Degree					
			Lt High School	High School	Junior College	Bachelor	Graduate	Total
Allow Homosexual to Teach	ALLOWED	Count	249	1270	216	584	326	2645
		% within RS HIGHEST DEGREE	68.4%	82.4%	91.9%	96.5%	97.0%	85.8%
	NOT ALLOWED	Count	115	272	19	21	10	437
		% within RS HIGHEST DEGREE	31.6%	17.6%	8.1%	3.5%	3.0%	14.2%
Total		Count	364	1542	235	605	336	3082
		% within RS HIGHEST DEGREE	100.0%	100.0%	100.0%	100.0%	100.0%	100.0%

TABLE 8

Chi-Square Tests			
	Value	df	Asymp. Sig. (2-sided)
Pearson Chi-Square	209.652[a]	4	.000
Likelihood Ratio	209.779	4	.000
Linear-by-Linear Association	167.021	1	.000
N of Valid Cases	3092		

their approval. Only 3.5% and 3% from the "educated" groups opposed gays' rights to teach.

These results beg the question whether the two variables are dependent or not? Moreover how strongly or weakly are they associated? The Null Hypothesis states that the variables are independent. In order to reject the latter, we will have to look up the significance level illustrated in the CHI-SQUARE TESTS table (Table 8).

The exact probability value Sig. (2-sided) equals 0.000, well below our alpha level of 0.05. Consequently, we reject the Null Hypothesis and state that there is a significant relationship between the variable "education" and the variable "colhomo". In addition to a significant relationship, the level of association of the variables is moderate. The Cramer's V in Table 9 (SYMMETRIC MEASURES) reveals a value between 0.11 and

Statistical significance is more important than the strength of the relationship

Phi: when you 2×2 matrix of variables

Cramers V: when at least 1 variabl has more than 2 categories

TABLE 9

Symmetric Measures		Value	Approx. Sig.
Nominal by Nominal	Phi	.260	.000
	Cramer's V	.260	.000
N of Valid Cases		3092	

TABLE 10

Favor or Oppose Gun Permits		Frequency	Percent	Valid Percent	Cumulative Percent
Valid	FAVOR	2318	48.1	74.7	74.7
	OPPOSE	785	16.3	25.3	100.0
	Total	3103	64.4	100.0	
Missing	IAP	1669	34.6		
	DK	40	.8		
	NA	8	.2		
	Total	1717	35.6		
Total		4820	100.0		

0.30, thus allowing us to conclude that the relationship is significant with a moderate strength.

Exercise 3

Over the past two decades, the American society has witnessed a dramatic increase in gun-inflicted crimes that culminated in 2012 in the tragic Newtown Sandy Hook shootings. At the same time, advocates of stricter gun control continue to lose ground to more organized gun lobbies.

In this exercise, we would like to see whether women and men differ on gun control issues. Are the variables "sex" (Respondent's Gender) and the variable "gunlaw" (Favor or Oppose Gun Permits) dependent or independent?

The FREQUENCIES table illustrates the number and percentage of respondents who have indicated "Opposing" or "Favoring" gun permits. A total of 25.3% have opposed gun permits and most likely would wish for looser gun control legislation. In contrast, 74.7% (or 2,018) respondents were in favor of the right to hold guns with the appropriate permit.

Thus, we are interested to see whether women on average would tend to favor gun permits, while men would oppose them. In order to carry out a Chi-Square Test, we click on ANALYZE -> DESCRIPTIVE STATISTICS -> CROSSTABS, and the CROSSTABS window will open.

CROSSTABS

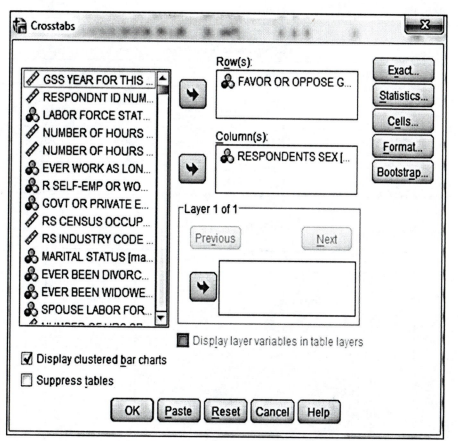

In the CROSSTABS window, we select "sex" as our independent variable (COLUMN(S)) and the variable "gunlaw", capturing respondents' sensitivity towards gun laws, as our dependent. Tables 11 and 12 illustrate the Chi-Square Test results.

The CROSSTABULATION in Table 11 displays the frequency and column percentages for male and female respondents. Surprisingly, the overwhelming majority of women (80.9% as opposed to solely 19.1%) expressed their support for gun permits. In contrast, solely 66.8% of men agreed to favor the latter, and 33.3% disagreed with any form of registration or licensing requirement prior to purchasing a gun. Consequently, women on average showed a rule-abiding demeanor and expressed their support for gun permits, while one-third of male respondents opposed such requirements. See the Crosstabulation output illustrated in Graph 2.

TABLE 11

Favor or Oppose Gun Permits * Respondents Sex Crosstabulation					
			Respondents Sex		
			Male	**Female**	**Total**
Favor or Oppose Gun Permits	FAVOR	Count	911	1407	2318
		% within RESPONDENTS SEX	66.8%	80.9%	74.7%
	OPPOSE	Count	452	333	785
		% within RESPONDENTS SEX	33.2%	19.1%	25.3%
Total		Count	1363	1740	3103
		% within RESPONDENTS SEX	100.0%	100.0%	100.0%

GRAPH 2

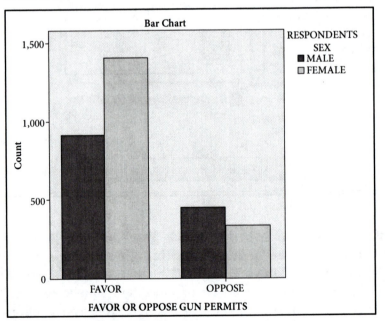

Is this relationship between the two variables significant? The Sig. (2-sided) value in the CHI-SQUARE TEST table indicates a significance of 0.000, lower than our alpha level of 0.05, which allows us to reject the Null Hypothesis. The latter states that the two variables are independent, and the relationship we could observe in the Crosstabulation table was caused by random chance alone. However, this likelihood is extremely

TABLE 12

Chi-Square Tests					
	Value	**df**	**Asymp. Sig. (2-sided)**	**Exact Sig. (2-sided)**	**Exact Sig. (1-sided)**
Pearson Chi-Square	79.543[a]	1	.000		
Continuity Correction[b]	78.802	1	.000		
Likelihood Ratio	79.150	1	.000		
Fisher's Exact Test				.000	.000
Linear-by-Linear Association	79.517	1	.000		
N of Valid Cases	3103				

TABLE 13

		Value	**Approx. Sig.**
Nominal by Nominal	Phi	−.160	.000
	Cramer's V	.160	.000
N of Valid Cases		3103	

low (0.000), and we were able to demonstrate that the relationship between the variables is significant.

The Measures of Association are reported in Table 13. Both our dependent and independent variables had two categories (2 × 2 matrix). Therefore, the correct Measure of Association students will need to look up is the Phi value. The latter indicated that while the relationship between the variable "sex" and one's attitude toward gun permits is significant (Chi-Square Test result), the association (or strength of relationship) between the two variables is weak.

higher the slope, the more responsive the dependent is
to changes in the independent.

$$Y = a + bx$$

b is slope
X is score on the
independent variable.

Y-intercept

score
on the
independent
variable

CHAPTER 9

Regression Analysis: Simple and Multiple Linear Regression

Chapter 9 will begin with a general overview of *regression analysis* and present five SPSS exercises introducing concepts such as *coefficient of determination, correlation coefficient, slope, partial slopes, Y intercept, beta weights, coefficient of multiple determination,* and *multiple correlation coefficient.*

In the first exercise, we will examine the effect of respondents' number of children on the number of household work hours. In Exercise 2, we would like to see if the effect of respondents' number of children on the number of their household activity hours increases, by considering only the sample of female spouses. More particularly, we are interested to see how the regression line changes if we look at how many hours female partners devote to household work given an increasing number of children. In the third exercise, we would like to examine the effects of education on respondents' total income. The last two exercises will test the effect of two independent variables ("degree" and "sex") on respondents' income. It addresses the question of whether the relationship between one's earning potential and the two independent variables capturing respondents' level of education and gender is statistically significant? The technique referred to as *multiple regression analysis* allows us to include more than one independent variable when probing a causal relationship. In the last exercise, we will carry out another *multiple regression analysis* for the purpose of examining the effects of education and race on respondents' income.

CONDUCTING A REGRESSION ANALYSIS IN SPSS

Exercise 1

The exercises and illustrations below use the 2012 GSS data set.

Before carrying out a regression analysis, students are encouraged to examine the relationship between two interval-ratio variables with the help of a scattergram. By selecting GRAPHS -> LEGACY DIALOGS -> SCATTER/DOT, the SCATTER/DOT dialogue box presents the researcher with five options.

SCATTERGRAM

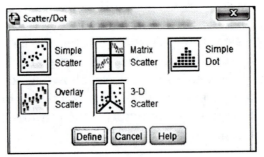

Students should click on SIMPLE SCATTER, followed by DEFINE.

SCATTER/DOT

After selecting the SIMPLE SCATTER option, the SIMPLE SCATTERPLOT dialogue box will open, allowing the researcher to select the dependent (Y axis) and independent (X axis) variables.

In this exercise we are interested to examine the effect of respondents' number of children (independent variable "childs") on the number of hours one devotes weekly to household work (dependent variable "rhhwork"). Thus, we select both variables from the variable list on the left and transfer the dependent variable to the Y AXIS box. Similarly, by clicking on the arrow sign, we move the independent variable "rhhwork" to the X AXIS box. After clicking OK, the scattergram illustrated in Graph 1 will appear.

SIMPLE SCATTERPLOT

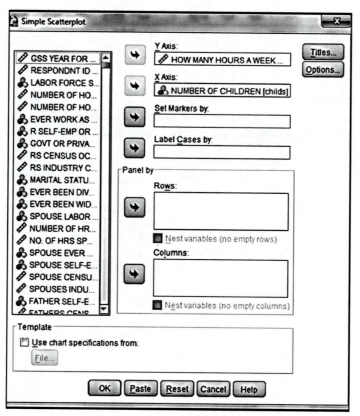

GRAPH 1

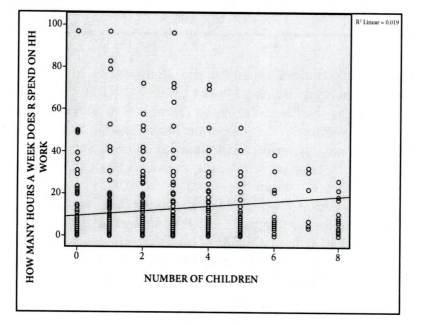

The scattergram in Graph 1 shows the independent variable, "childs", on the X axis and the dependent variable, "rhhwork", on the Y axis. In order to better display the relationship between the independent and the dependent variable, we double-click on the scattergram, and the CHART EDITOR window will open.

CHART EDITOR

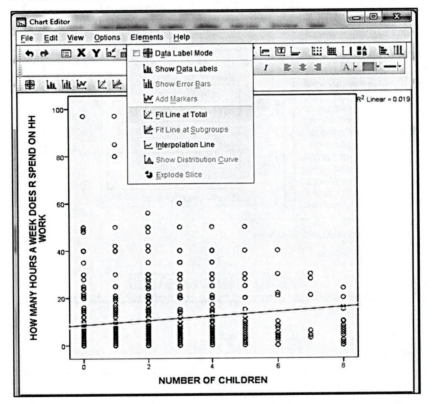

By clicking on the ELEMENTS menu option, followed by FIT LINE AT TOTAL, a least-squares regression line, illustrated in the CHART EDITOR table, will be added to the original scattergram. The latter allows the researcher to better predict the relationship between the two variables. The regression line's positive trend reveals that as the number of children increases, respondents tend to devote a larger amount of time to manage household work. In other words, the underlying assumption that "more kids equal more household work" cannot be refuted.

However, due to numerous overlapping data points (i.e., respondents scoring similarly on both variables) the scattergram's capacity to accurately describe the relationship has been weakened by the inclusion of all observations. It is recommended that students select only a small subset of cases to exemplify the two variables' relationship. Thus, we click on DATA -> SELECT CASES and choose a RANDOM SAMPLE OF CASES. (See below.)

SELECT CASES: RANDOM SAMPLE

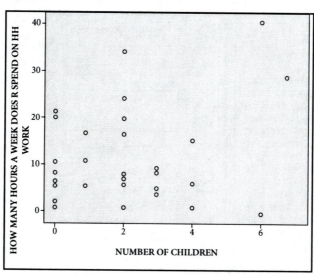

Students can manually enter a percentage from 1% to 5%, depending on the size of the data set. Following the selection of a smaller random sample, we click again on GRAPHS -> LEGACY DIALOGUES and SCATTER/DOTS. The following output illustrated in Graph 2 will be produced.

GRAPH 2

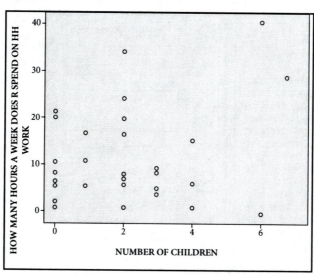

Regression requires an interval ratio dependent

The scattergram indicates that the relationship between the two variables is positive (i.e., when one variable is increasing the other variable will increase as well) and linear.

The two cases illustrated at the upper right corner reflect the experiences of two respondents from our random sample. Both indicated to have more than five children and, at the same time, to spend from 30 to 40 hours weekly on household work. In contrast, respondents who have fewer children (i.e., one or two) tend to spend fewer hours on household tasks.

Following the scattergram illustration, the researcher is now interested to calculate the linear regression line's slope (b) and Y intercept (a), as well as the coefficient of correlation (Pearson's R) and the coefficient of determination (R^2).

It is important for students to remember that the IF CONDITION is a "sticky" factor that should be deactivated before taking the next steps to other calculations. Click on DATA -> SELECT CASES and select the ALL CASES option in order to include each and every case.

Following the IF CONDITION's deactivation, we click on ANALYZE -> REGRESSION -> LINEAR, and the LINEAR REGRESSION window (illustrated below) will open. The latter allows students to select the dependent and independent variables from the variable list located at the left. We select the variable "rhhwork", capturing the number of hours respondents spend on household chores, as our dependent variable (the effect/result), and the variable "childs" (respondents' number of children), as our independent. The two interval-ratio variables revealed a positive linear relationship. (See scattergram.)

However, the researcher also wants to compute the regression line's components (i.e., the slope and Y intercept) as well as the strength of the relationship (i.e., Pearson's R) and the percentage of variance explained in the dependent variable by the independent (i.e., Coefficient of Determination).

LINEAR REGRESSION

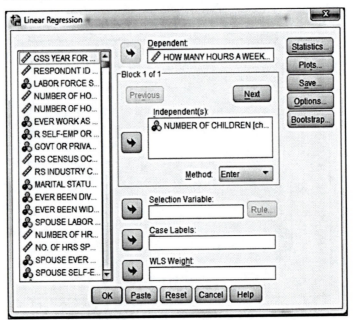

TABLE 1

Model Summary				
Model	**R**	**R Square**	**Adjusted R Square**	**Std. Error of the Estimate**
1	.139[a]	.019	.018	11.352

a. Predictors: (Constant), NUMBER OF CHILDREN

After clicking on OK, the MODEL SUMMARY, ANOVA, and COEFFICIENTS tables will be generated.

In Table 1 we will find the values of Pearson's R and the R Square. The Pearson's R, or the *correlation coefficient*, captures the association between variables measured at the interval-ratio level. The value of Pearson's R varies from -1 to 1, where "-1" means a perfect negative correlation while "+1" a perfect positive correlation. In other words, a Pearson's R equal to 1 indicates that a one-unit increase in the independent variable triggers a parallel one-unit increase in the dependent. In contrast, a negative "-1" value reveals that a one-unit increase in the independent will cause a similar one-unit decrease in the dependent variable. While the correlation coefficient encompasses values from -1 to 1, a value of zero would reflect no association. For values other than the ones described above, we would find associations:

- between -0.30 and 0 or 0 and 0.30 to reflect a weak relationship
- between -0.60 and -0.31 or 0.31 and 0.60 to reflect a moderate relationship
- between -1 and -0.60 or 0.60 and 1 to reflect a strong relationship

In addition to the categories listed above, it is important for students to remember that a negative value for the correlation coefficient reveals a negative association, while a positive value a positive relationship.

In this example, a Pearson's R value of 0.139 describes a positive but weak relationship. That is, while the independent increases (i.e., number of children increases), the number of hours one spends on household work will increase as well. However, the latter (e.g., dependent variable) is affected only weakly by the increase in the values of the independent.

In addition to the Pearson's R, Table 1 also illustrates the value for R Square, or the *coefficient of determination*. In contrast to Pearson's R, the coefficient of determination, similarly to Lambda, works according to the Proportional Reduction in Error (PRE) logic. In other words, PRE measures seek to predict the dependent variable's value, first by disregarding the information supplied by our independent and later on accounting for the latter's importance in predicting the dependent's scores. A value of 0.019, illustrated in Table 1, reveals that the independent variable (i.e., number of children) explains only 1.9% of the total variation in the number of hours respondents devote to household tasks (our dependent). Thus, approximately 98.1% of the variation in the dependent variable is left unexplained by our independent.

TABLE 2

ANOVA[a]					
Model	Sum of Squares	df	Mean Square	F	Sig.
1 Regression	3218.786	1	3218.786	24.975	.000[b]
Residual	164190.414	1274	128.878		
Total	167409.200	1275			

a. Dependent Variable: HOW MANY HOURS A WEEK DOES R SPEND ON HH WORK
b. Predictors: (Constant), NUMBER OF CHILDREN

TABLE 3

Coefficients[a]		Unstandardized Coefficients		Standardized Coefficients		
Model		B	Std. Error	Beta	t	Sig.
1	(Constant)	8.510	.471		18.086	.000
	NUMBER OF CHILDREN	.940	.188	.139	4.998	.000

a. Dependent Variable: HOW MANY HOURS A WEEK DOES R SPEND ON HH WORK

The ANOVA table (Table 2) displays the significance of the relationship between the independent variable and the dependent. The Sig. value equals 0.000, less than 0.05, thus we can reject the Null Hypothesis. The H_0 assumed that the variables' relationship is not statistically significant and is attributable to random chance alone. We conclude that the relationship between number of children and number of hours spent weekly on household work is significant, thus, rejecting the H_0.

The COEFFICIENTS table (Table 3), outlines the values for the slope and the Y intercept. The slope reveals by how much the dependent variable changes (increase or decrease) given a *one-unit* increase in the independent variable. Consequently, a slope of 0.940 reflects that for each additional child, the time devoted to household chores increases by approximately 1 hour. In contrast, the value of the Y intercept shows us the number of hours spent weekly on household work, while our independent variable equals zero—that is, when one doesn't have any children.

Thus, a value of 8.510 for our Y intercept displays that when the number of children is zero, respondents still have to devote up to 8.5 hours to household activities. As a result of the information provided by Table 3, the regression line will take the following form: $Y=8.510+0.940X$.

This equation formula will help the researcher to predict the values for the dependent variable ("rhhwork") given any value for the independent ("childs"). Thus, if we would be curious to predict the number of hours Americans devote weekly to household work, assuming that they have six children, the equation would take the following form: Y=8.510+0.940*6. Consequently, the time allotted to household work would be equal to 14.15 hours.

Exercise 2

In Exercise 2, we would like to see whether the effect of respondents' number of children on the number of hours dedicated to household activities increases if we take into account solely the sample of female spouses. More particularly, we are interested to see how the regression line changes if we look at how many hours female partners devote to household work, given an increasing number of children. Or, in other words, how do more children impact the number of work hours at home?

The variable "sphhwork" (i.e., How many hours a week does spouse devote to household work?) captures the number of hours respondents' spouses dedicate to household chores. Since we are interested solely in female spouses, we will select only male respondents, assuming that heterosexual relationships are the norm in the 2012 GSS data set.

We click on DATA -> SELECT CASES -> IF CONDITION IS SATISFIED, and the following SELECT CASES: IF window will open.

Besides the "SEX=1" requirement that allows only for the selection of male respondents, we add an additional prerequisite. The variable "childs", capturing respondents' number of children, should be equal to or higher than 1; that is, the two conditions together will allow for the selection of male respondents (with female spouses) who have at least one child together. We have attached the latter condition for the purpose of conducting a more thorough test of the independent variable's effect on the dependent variable.

SELECT CASES: IF

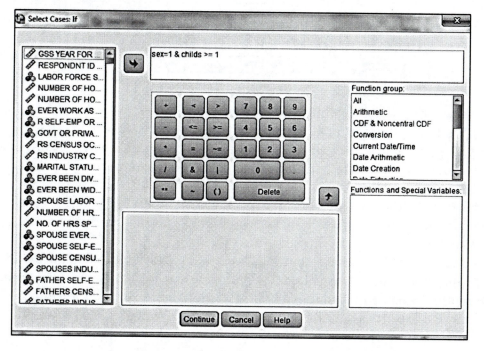

TABLE 4

Model Summary				
Model	R	R Square	Adjusted R Square	Std. Error of the Estimate
1	.175[a]	.031	.027	18.232

a. Predictors: (Constant), NUMBER OF CHILDREN

Following the IF CONDITION, we click on ANALYZE -> REGRESSION -> LINEAR, and the below illustrated LINEAR REGRESSION window will open.

LINEAR REGRESSION

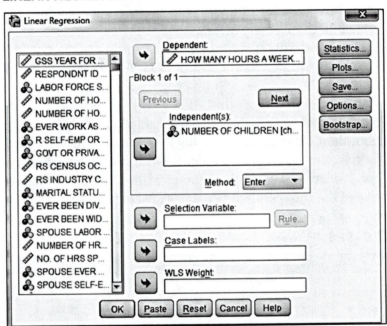

We select the variable "sphhwork" as our dependent and the variable "childs" as the independent. After clicking OK, SPSS will generate three output tables.

According to Table 4, the Pearson's R value (0.175) reveals a weak, positive association between the variables. In addition, the coefficient of determination tells us that the independent variable (number of children) explains only 3.1% of the total variation in the number of hours respondents' spouses devote to household chores (our dependent). Thus, approximately 96.9% of the variation in the dependent variable remains unexplained by our independent.

The ANOVA table displays the Sig. value and reveals whether the relationship between the independent variable and the dependent is significant. Once again, the Null Hypothesis states that the relationship is not significant and attributable solely to random chance. However, Table 5 points to a Sig. value of 0.005, less than 0.05,

TABLE 5

ANOVA[a]						
Model		Sum of Squares	df	Mean Square	F	Sig.
1	Regression	2709.111	1	2709.111	8.150	.005[b]
	Residual	85759.085	258	332.400		
	Total	88468.196	259			

a. Dependent Variable: HOW MANY HOURS A WEEK DOES SPOUSE ON HH WRK
b. Predictors: (Constant), NUMBER OF CHILDREN

TABLE 6

Coefficients[a]						
Model		Unstandardized Coefficients		Standardized Coefficients	t	Sig.
		B	Std. Error	Beta		
1	(Constant)	12.994	2.326		5.587	.000
	NUMBER OF CHILDREN	2.152	.754	.175	2.855	.005

a. Dependent Variable: HOW MANY HOURS A WEEK DOES SPOUSE ON HH WRK

allowing us to reject the Null Hypothesis. Consequently, we can conclude that there is a statistically significant relationship between the number of children one has and the number of hours one's spouse devotes to household work.

The COEFFICIENTS output block displays the Y intercept (Constant) and the value of the slope. A slope of 2.152 indicates an increase in the number of hours one's spouse devotes to household responsibilities, given a one-unit change in the independent. In other words, respondents' spouses will put in an extra 2.1 hours of household work with an additional child to take care of.

If we contrast this value with the slope illustrated in Exercise 1 (0.940), we find that the former (2.152) is much steeper. A steeper slope indicates that a one-unit change in our independent variable has a stronger effect on our dependent variable. In this case, an additional child would increase the respondents' spouses' hours of household work by 2.1 hours, while in the previous example, respondents would spend approximately one additional hour on household choirs parallel to a one-unit increase in our independent (i.e., number of children increases by one). Consequently, for women in our second sample, the impact of an additional child to hours devoted for household work is more than double than for respondents (both male and female) investigated in Exercise 1.

Similarly, the Y intercept reflects how many hours of work female spouses devote to household chores when the effect of the independent variable is not taken into account. Thus, when the number of children is zero, female spouses still spend up to 13 hours (12.994) on household tasks. In addition to a steeper slope, respondents' wives spend weekly an extra 4.5 hours (the difference between the Y intercept illustrated in Exercise 1 and the Y intercept calculated for Exercise 2) more on household work.

Finally, the simple linear regression line will take the following form: Y=a+bx—that is, Y=12.994+2.152X. The latter allows us to predict values for the dependent variable ("sphhwork") for any value of "childs" (our independent).

Exercise 3

In this exercise, we would like to examine the effects of education (independent variable "degree") on respondents' total income (dependent variable "realrinc"). While students should choose independent variables measured at the interval-ratio level, from time to time, they can replace the independent variable with correctly coded ordinal variables that encompass a broad range of scores.

The independent variable "degree" (Respondent's Highest Degree) contains five categories (Little High School, High School, Junior College, Bachelor, and Graduate Degree), and the variable's ascending coding correctly reflects the earned degrees' surging tendency. (See VALUE LABELS window.)

VALUE LABELS

In order to conduct the Regression Analysis, we click on ANALYZE -> REGRESSION -> LINEAR, and the LINEAR REGRESSION window will open.

TABLE 7

Model Summary				
Model	**R**	**R Square**	**Adjusted R Square**	**Std. Error of the Estimate**
1	.268[a]	.072	.072	53825.780

a. Predictors: (Constant), RS HIGHEST DEGREE

LINEAR REGRESSION

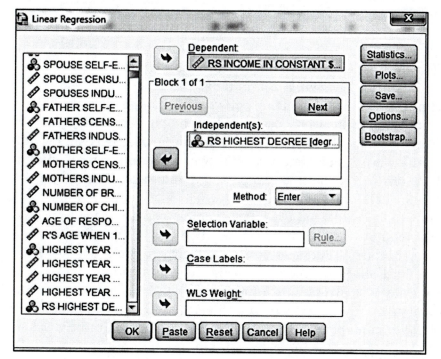

We move the dependent interval-ratio variable "realrinc", capturing respondents' income, into the DEPENDENT box, and the independent ordinal variable "degree" (Respondents' Highest Degree) into the INDEPENDENT(S) box. Following the selection of the variables, students should click on OK. Similarly to Exercise 1 and 2, three output tables will be generated.

The first output (MODEL SUMMARY) illustrates the values for the correlation coefficient (Pearson's R) and the coefficient of determination (R square). An R value of 0.268 indicates that the relationship between the two variables is positive and weak (or slightly moderate). While the relationship is weak, the positive trend reveals that, on average, higher levels of education lead to higher levels of income. In addition, the coefficient of determination shows us that 7.2% of the variation in income is explained by the level of education.

TABLE 8

Model		Sum of Squares	df	Mean Square	F	Sig.
ANOVA[a]						
	Regression	634145324423.999	1	634145324423.999	218.881	.000[b]
	Residual	8184631355711.772	2825	2897214639.190		
	Total	8818776680135.771	2826			

a. Dependent Variable: RS INCOME IN CONSTANT $
b. Predictors: (Constant), RS HIGHEST DEGREE

The Sig. value in Table 8 (0.000) confirms that the relationship between education and income is statistically significant. In other words, we can reject the Null Hypothesis, which posits that the relationship between the independent variable "degree" and the dependent variable "realrinc" occurred by random chance alone.

So far we know that the effect of education on income is positive and statistically significant with a weak/moderate strength. Nonetheless, the researcher is also interested in the linear regression line's slope and Y intercept (Constant). Both values are illustrated in the COEFFICIENTS table.

The slope (B) in Table 9 indicates that income increases on average by $12,151.42 concurrently with an increment by one unit in the independent variable. In other words, as we advance from one education category to the next level (i.e., from High School to Junior College Degree or from Bachelor to Graduate Degree), income will rise on average by $12,151.42. It is important for students to remember that the slope always measures *the effect of a one-unit increase* in the independent on the dependent variable. If the independent variable is measured on an ordinal scale, *a one-unit increase in the independent* indicates a jump by one unit from one category to the subsequent category. In contrast, a Y intercept of $5,699.52 indicates the value of the dependent variable when respondent's education level is zero. From the information illustrated in Table 9, the regression equation will take the following form: $Y=5699.52+12151.426X$.

Exercise 4

In the last two exercises, we will examine the effects of *two* independent variables ("degree" and "sex") on respondents' income (dependent variable "realrinc"). The technique referred to as *multiple regression analysis* allows us to include more than one independent variable when probing a causal relationship.

We click on ANALYZE -> REGRESSION -> LINEAR, and the below illustrated LINEAR REGRESSION window will be generated.

LINEAR REGRESSION

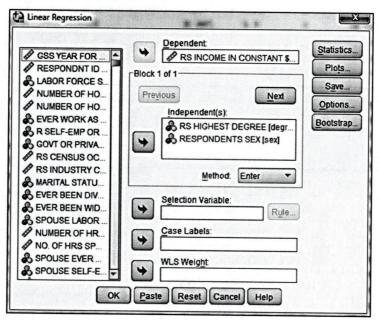

Similarly to Exercises 1–3, the researcher specifies the dependent and independent variables he/she has chosen for the multiple regression analysis in the LINEAR REGRESSION window. Thus, we move the dependent variable "realrinc" into the DEPENDENT box and the two independent variables (Respondent's Highest Degree and Respondent's Sex) into the INDEPENDENT(S) box. In contrast to simple regression analysis, we select *more than one* independent variable for the INDEPENDENT(S) box.

As discussed in Exercise 3, we are allowed to test the effects of the ordinal variable "degree" on one's earning potential as long as the independent variable encompasses a broad range of correctly coded categories (i.e., reflecting a higher value in a given category compared to the previous). In some instances, variables that contain only two categories (such as Gender) can also be included among the list of independents. Consequently, we move the variables "sex" and "degree" to the INDEPENDENT(S) box. After we click OK, the MODEL SUMMARY, ANOVA, and COEFFICIENTS tables will be generated.

Table 10 illustrates the values for the *multiple correlation coefficient* (0.311) and the *coefficient of multiple determination* (0.097).

In contrast to the correlation coefficient (Pearson's R) discussed in simple regression analysis, the multiple correlation coefficient indicates the strength of the relationship between the dependent variable and our independent variables (such as two or more independent variables). Consequently, a value of 0.311 reveals a moderately strong relationship between our two independent variables ("degree" and "sex") and our dependent ("realrinc").

TABLE 10

Model Summary				
Model	R	R Square	Adjusted R Square	Std. Error of the Estimate
1	.311[a]	.097	.096	53117.015

a. Predictors: (Constant), RESPONDENTS SEX, RS HIGHEST DEGREE

TABLE 11

ANOVA[a]						
Model		Sum of Squares	df	Mean Square	F	Sig.
1	Regression	851094196744.850	2	425547098372.425	150.827	.000[b]
	Residual	7967682483390.922	2824	2821417309.983		
	Total	8818776680135.771	2826			

a. Dependent Variable: RS INCOME IN CONSTANT $
b. Predictors: (Constant), RESPONDENTS SEX, RS HIGHEST DEGREE

Similarly, the coefficient of multiple determination (R Square) reveals the extent to which all independent variables collectively explain the variation in the dependent. According to Table 10, a coefficient of multiple determination of 0.097 indicates that 9.7% of the variation in respondents' income is jointly explained by education and gender.

The ANOVA table displays a Sig. value of 0.000, lower than 0.05, which prompts us to reject the Null Hypothesis. The H_0 stated that the relationship between the dependent variable and the two independent variables is not significant and was caused by random chance alone. We reject the latter and conclude that the relationship between one's earning potential and the two independent variables capturing respondent's level of education and gender is statistically significant.

The COEFFICIENTS table (Table 12) displays the values for both *partial slopes*, the Y intercept (Constant), and the *beta weights*.

A Y intercept reveals one's earning potential ($31,941.083) if we do not take the two independent variables into account. Since we have two independent variables, Table 12 illustrates two additional partial slope values: (1) a slope for education and (2) a slope for gender.

The slope for education indicates that income increases by $12,384.896 on average as we move up from one education category to the next (i.e., from a Junior College to a Bachelor Degree), *while controlling for gender*. The implication of "while controlling for gender" is quite significant. In contrast to simple regression analysis, multiple regressions contain two or more independent variables, while each variable's partial slope measures the former's impact on the dependent variable. However, this impact

TABLE 12

		Unstandardized Coefficients		Standardized Coefficients		
Model		B	Std. Error	Beta	t	Sig.
1	(Constant)	31941.083	3482.861		9.171	.000
	RS HIGHEST DEGREE	12384.896	810.962	.273	15.272	.000
	RESPONDENTS SEX	−17543.943	2000.701	−.157	−8.769	.000

Coefficients[a]

a. Dependent Variable: RS INCOME IN CONSTANT $

can be calculated only if we can control for the influence of all other independent variables.

Thus, the partial slope for gender reveals the change in income (i.e., increase or decrease) given a *one-unit change* in the independent ("sex"), while *controlling for education*. In other words, the illustrated negative value stands for a $17,543.943 decrease in income, given a one-unit change in gender, while controlling for education. In order to understand what we mean by a "one-unit change in gender", we have to revisit the coding of the variable "sex". (See below.)

VALUE LABELS

According to the VALUE LABELS window, female respondents received a code of "2" while men a code of "1". Consequently, a one-unit change for the variable "sex" means that we are moving from the first category ("1=MALE") to the second category ("2=FEMALE"). In other words, the partial slope captures an income decrease in the amount of $17,543.943, while we control for education.

Consequently if we contrast the two partial slope values, gender's impact on one's earning potential is more substantial than the increase caused by a higher level of education. The latter result is disconcerting and reveals a gender gap in which a respondent's sex has a greater influence on one's income than his/her level of education. Since both variables are measured in different units of measurement ("sex" encompassing two categories while "degree" five), comparing the absolute value of their respective slopes would make us draw incorrect conclusions. In such instances, where we would like to compare the independents' impact on our dependent, we will have to look up the *beta weights* (or *standardized coefficient*) values illustrated in Table 12. SPSS converts all values into Z scores (referred to as *standardization*) and recalculates partial slopes for each standardized independent variable. Consequently, the beta weights allow us to compare the effect of each of the independent variables on the dependent variable, while controlling for the remaining independent(s). In this example, a beta weigh of 0.273 for education, in contrast to a value of 0.157 for gender, reveals that education's impact on one's income is stronger than his/her gender. Thus, the regression line will take the following form: $Y=31941.083+12384.896X_1-17543.943X_2$, allowing us to predict future values for "realrinc" given any value for "degree" and "sex".

Exercise 5

In this last exercise, we will carry out another multiple regression analysis where we will test the effects of education and race (independent variables "degree" and "race") on respondents' income (dependent variable "realrinc").

Click on ANALYZE -> REGRESSION -> LINEAR, and the LINEAR REGRESSION window will appear. As illustrated below, variables "degree" and "race" are assigned to the INDEPENDENT(S) box while "realrinc" to the DEPENDENT box.

LINEAR REGRESSION

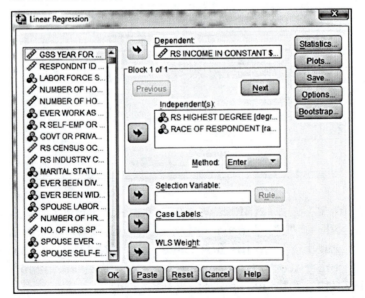

TABLE 13

Model Summary				
Model	R	R Square	Adjusted R Square	Std. Error of the Estimate
1	.271[a]	.073	.073	53790.370

a. Predictors: (Constant), RACE OF RESPONDENT, RS HIGHEST DEGREE

TABLE 14

ANOVA[a]						
Model		Sum of Squares	df	Mean Square	F	Sig.
	Regression	647804069239.537	2	323902034619.768	111.945	.000[b]
1	Residual	8170972610896.235	2824	2893403899.043		
	Total	8818776680135.771	2826			

a. Dependent Variable: RS INCOME IN CONSTANT $
b. Predictors: (Constant), RACE OF RESPONDENT, RS HIGHEST DEGREE

Similarly to Exercise 4, Table 13 illustrates the value for the multiple correlation coefficient (Pearson's R) and the coefficient of multiple determination (R Square). A value of 0.271 for Pearson's R indicates a weak or slightly moderate relationship between the dependent (income) and our independent variables (education and race). On the other hand, an R square value of 0.073 shows that one's level of education and his/her race taken together explain 7.3% of the variation in income.

According to the ANOVA table, the relationship between income, education, and race is statistically significant. The Sig. value illustrated in Table 14 allows us to reject the Null Hypothesis.

The COEFFICIENTS table displays the value for the Y intercept (Constant), as well as the two partial slopes for each independent variable. To make the comparison between the standardized partial slopes possible, Table 15 also contains the beta weights (or standardized coefficients).

The value of the Y intercept is $10,575.322, and the partial slope for education is $12,010.241, which means that income will increase by the latter amount if respondent's obtained degree moves up by one level, while controlling for race. Similarly, the partial slope for race illustrates that income will decrease by approximately $3,481 on average, if the *independent changes by one unit*, while controlling for education. In order to understand what such "change" in the independent might entail, it is important to reexamine the coding of the independent variable "race".

TABLE 15

	Unstandardized Coefficients		Standardized Coefficients		
Model	**B**	**Std. Error**	**Beta**	**t**	**Sig.**
(Constant)	10575.322	2879.564		3.673	.000
1 RS HIGHEST DEGREE	12010.241	823.368	.265	14.587	.000
RACE OF RESPONDENT	−3481.958	1602.591	−.039	−2.173	.030

Coefficients[a]

a. Dependent Variable: RS INCOME IN CONSTANT $

VALUE LABELS

The VALUE LABELS window illustrates that the variable "race" has three categories ("1=WHITE", "2=BLACK" and "3=OTHER"). Consequently, a one-unit change in the given independent variable would mean that we are moving from the category of Caucasian respondents to the sample of African-Americans. As we switch from 1 to 2, income decreases by $3,481 on average, while controlling for education. If we contrast the latter with the partial slope for gender in the previous exercise, we can see that moving from "male" to "female" had a more detrimental effect on income than a one-unit change in race. (i.e., moving from "white" to "black" or from "black" to "other"). Moreover, the beta weights illustrated in Table 15 demonstrate that education's impact on income is stronger than the respondent's race. The equation for the multiple regression line will take the following form: $Y=10575.322+12010X_1-3481.958X_2$.

Index